The Calm Bi

The practical guid
mamas to create a calm, positive
hypnobirth

By

Suzy Ashworth

Copyright 2015 Suzy Ashworth, The Calm Birth School Ltd. All Rights Reserved.

No part of this book may be produced by any means, nor transmitted, nor translated into a machine language without written permission of the publisher.

The right of Suzy Ashworth to be identified as the author of this work has been asserted in accordance with sections 77 and 78 of the Copyright, Designs & Patent Act 1988.

Conditions of Sale. This book is sold subject to the condition that it shall not, by way of trade 3 or otherwise be lent, resold, hired out or otherwise circulated with the publisher's prior consent in any form of binding or cover other than that in which it is published and without a similar condition, including this condition, being imposed on the subsequent purchaser.

First published by Suzy Ashworth in Great Britain in 2015.

ISBN: 978-1-326-53117-1

DISCLAIMER: The techniques are designed to provide helpful information on pregnancy and birth. The author is not medically or clinically trained and the content of this book is in no way intended to be represented as medical advice nor as an alternative to appropriate medical procedure. The reader should seek the advice of a medical doctor to answer any health-related issues, questions or concerns she may have surrounding her pregnancy, labour or the delivery of her baby. The advice shared in this book is in no way a guarantee or promise of expected, imagined or actual outcome of labour or birth. Neither the author nor the publisher accept responsibility for any liability or damage caused by or as a result of practicing the techniques in this book.

Dedication

I dedicate this book to Jerome, Caesar, Coco and The Bump.
All the love. xxx

Contents

Introduction: No Vagina Whispering

1. How to Use This Book
- How to Use This Book
- Laura's Story

2. What is Hypnobirthing?
- How Do You Create Your Own Calm Birth?
- The Evolution of Birth
- The Biology of the Birthing Body
- The Neuroscience Bit

3. An Introduction to Relaxation
- Reasons to Relax
- Hypnosis/Trance
- The Brain
- The Calm Birth School Breathing Technique
- Emiliana's Story

4. The Biology Bit
- The Uterus and What Happens in Labour
- The Uterus in Surge
- Why Endorphins Are Your New Best Friend
- Pain
- Karis' Story
- Wave Breathing

5. Taking Control

- Doing the Work
- Language
- The Calm Birth School Glossary
- Birthing Horror Stories
- Release and Let Go

6. Where to Give Birth
- Creating a Positive Birth Environment
- Step 1: Know Your Options
- Katie's Story
- Step 2: Start Asking Questions and Investigate
- Step 3: Get to Know Your Environment
- Step 4: Make it Your Own

7. Choosing Your Care Providers
- Things to Consider
- Who Is Going to Be in the Room With You?
- Holding the Space

8. Communicating With Your Care Providers
- Get Specific
- Use Your BRAIN
- Birth Preferences
- Ana's Story

9. Understanding Estimated Due Dates
- Think 'Guess Date' Not 'Due Date'

10. Preparing For Labour and Birth
- Pelvic Floor Muscles
- Perineal Massage
- Prenatal Bonding

- Hazel's Story

11. Your Birth Partner
- The Birth Partner's Role
- During Pregnancy
- On the Day
- Alex's Story

12. The Induction Process and Special Circumstances
- Continuous Monitoring
- Pre-Labour Interventions
- The Induction Process
- Special Circumstances

13. How to Help Your Baby Along Naturally
- Nudges in the Right Direction
- Emma's Story

14. What to Do When Labour Starts
- The Seven Signs of Early Labour
- During the Night
- During the Day
- Calm Birth School Recommendations for Early Labour
- Natasha's Story

15. Positions for labour
- Active Birthing: Sitting/Squatting, or on Your Hands and Knees
- Keeping Mobile During Active Labour
- Irene's Story

16. Active or Established Labour
- What Active Labour Feels Like
- When to Travel to Your Care Provider

17. Coping With Distractions
- How to Deal With Distractions
- Massage
- The Dial Down Method
- Time Distortion
- Laura's Story

18. What Happens When You're With Your Care Provider
- A Moment in Time
- The Slow Down
- Working With Your Care Provider During Labour
- Language

19. Second Stage Labour
- A Pause
- Pooing
- Feeling Nauseous
- Pushing and Breathing
- Birth Breathing

20: Transition and Crowning
- I Can't Do This!
- Charlotte's Story

21. Your Baby's Here!
- Cleaning Your Baby

- Delayed Cord Clamping
- Birthing Your Placenta
- Skin to Skin
- Breastfeeding

Conclusion

Post Script
- Hayley's Story

Acknowledgements

Further Reading

Introduction:
No Vagina Whispering

So you've finally bitten the bullet, or maybe some kind soul bit the bullet for you and has kindly introduced you to The Calm Birth School, founders of the world's first video-based hypnobirthing program. Since our launch in December 2014 the Calm Birth School family has helped three thousand women across six different continents create calm and positive birth experiences, so you are in *the* best company. But enough about us, let's cut to the chase and acknowledge that you – yes, you – are right now growing another human being, even as you read these words. And that is bar far the biggest deal in the world. You are bloody amazing.

Perhaps the whole 'growing a baby' thing is something that you have already got to grips with. If not, go and Google it. There are a million and one blogs and forums telling you exactly what size nut your baby is today. This book isn't about any of that though. This book is about the most amazing feat a woman's body goes through: birth.

It's also about far more than "just" birth. Yes, you'll get the full lowdown on the different tips, tricks and techniques that provide the foundation for thousands of women creating empowering, positive and fear-free birth experiences every day. But alongside being a book about birth, this is a journey

where you will learn how to unleash your personal power from within, so that when the moment finally arrives for you to welcome your baby into the world, you feel calm, positive and ready for whatever twists and turns may be presented to you. The one thing we know for sure about birth is that it's a journey, the type where you can't know ahead of time exactly what's going to happen. You can however, arm yourself with a mindset and a toolkit that can make giving birth one of the most incredible and enjoyable experiences of your life.

By the time you have finished leafing through these pages you will understand exactly what you need to do to give yourself permission to go wherever you need to go, controlling what you can control and letting go of what you can't as you eliminate the fear of birth. It's this that enables you to do what you need to in order to bring your baby into the world positively, happily and confidently.

Am I serious about birth being happy, positive and enjoyable? Deadly. Am I a lentil-eating, hummus-knitting, crochet-knickers-wearing kind of woman? Absolutely not. Am I going to make you walk from room to room, wafting joss sticks around while chanting in Sanskrit? Will I be asking you to make a homage to the lotus flower or whisper to your vagina? No, no and no: there shall be no vagina whispering! (Unless of course, you want there to be.)

I am not a hippy. I write these words to you as an everyday woman who has been there and got the calm birth t-shirt not once but twice. I'm here to show you that no matter who you are or what your perspective on birth is, your birth can be an experience you will never forget, for all the right reasons.

While you might be feeling confident in your body's ability to do this job (and why the hell shouldn't you be? This is after all quite literally what your body is designed to do), it's also highly possible that you are peeing your metaphorical pants at the thought of everything to do with the bit before you hold your baby in your arms for the first time.

If this is your first child, and if you're anything like me or many of the clients I work with on a daily basis, you might be in a bit of a head spin with the realisation that there is an actual baby growing inside you. That's okay: it's normal to get a bit freaked out as you comprehend that this collection of cells that started out as smaller than a full stop will one day become an actual human being.

Or perhaps the mere thought of squeezing a nine pounder out of your lady parts brings tears to your eyes... so you aren't thinking about it. Maybe you are so busy being Superwoman that the fact you're about to become a mother *and* give birth has barely registered on your inner Richter Scale. You're just too goddamn busy right now to really give it any head space. Or, maybe you're a control freak who needs information, and lots of it. For you, the thought of your body doing things you can barely bring yourself to imagine makes you want to reach for a large gin, so you're arming yourself with as much information as you can get. As I have said before, I've been there and done this, and I work with women like you every day. I want you to know that it doesn't matter who you are or whether you're experienced or knowledgeable about birth: you really *have* got this.

By reading this book, you're acknowledging that at least a small part of you knows that birth doesn't have to be like the horror stories you've seen on TV, discussed with girlfriends or overheard around the water cooler. You don't have to be a disempowered woman who feels scared, vulnerable and as though her birth is happening *to* her. You don't have to go into birth unaware of all of the things you don't know just because you haven't done this before.

The Calm Birth School's mission is to arm you with knowledge and empower you with confidence, so that you know from the very core of your being that in being a woman, you were *designed* to do this. I am here to remind you that the innate ability to give birth, genetic programmed in womankind over hundreds of thousands of years, is right at your fingertips.

Will this guarantee you an easy, straightforward birth where your baby pops out reciting the alphabet backwards? (Hypnobabies are very advanced you know.) Sadly no. I wish I could guarantee such linguistic feats alongside a "perfect" birth, but that isn't the case. So what is the point of hypnobirthing then?

I'd like to invite you to think of your birth as the most important meeting of your life to date. What would you do before you knocked on the office door, ready to strut your stuff in front of your boss? You would prepare everything down to the nth degree. You would have every base covered so that whatever curveballs you may be faced with, you'd know you were able to handle it all. You'd feel calm, confident and ready to go.

You already know that as first meetings go, the date you have with your baby is going to be infinitely more impactful, meaningful and enjoyable than any meeting with a boss. My goal is for you to feel even more confident, prepared and knowledgeable than you would be if you were prepared to come face to face with the people who pay your wages.

Why is this so important? Well, contrary to popular opinion, I will unreservedly declare that giving birth should be one of the most empowering, life-affirming and joyful events you will ever experience. When you learn to work *with* your body, as opposed to against it, you will know how to embrace the sometimes weird, sometimes wonderful power and intensity that moves up, down, around and through your body. And when you allow yourself to go with that flow, knowing with every fibre of your being that you were born to do this, then yes, it really can feel amazing.

Rachel's Story

Yesterday my darling baby girl Wynter arrived and I want to tell you about my and Wynnie's calm birth experience. I was overjoyed to have achieved the birth I had dreamed of! Seeing as she is my first, I didn't expect things to progress as quickly as they did, but after my waters released at 1:45am, I was in the midwife-led birthing unit at The Princess Alexandra Hospital by 5:00am where I was 3cm dilated.

Breathing and TENS took me to 6/7cm and then I was able to get in

the pool for the final part of our journey, leading to her safe arrival at 10:45am. My husband, who was an excellent birth partner, kept me strong enough to avoid any pain relief that would pass the placenta or let me lose focus. He said he'd never seen me like that before. I was even sleeping between contractions! In turn, baby's heart rate was 'perfect' throughout the birth, even during my hour-long pushing stage.

I still feel relaxed now and baba's been sleeping so much since the birth; we have to keep waking her to make sure she feeds! All is very well and I'd like to thank you for helping me achieve the birth I wanted.

The calm birth experience is for you, of course, but it's also for your baby too; when you release the resistance that impedes labour and birth, your baby's experience of entering the world becomes infinitely less jarring. A calm and positive end to your pregnancy and start to the newest chapter of your family's life also helps you focus on the most important job of all: learning how to be mum to this new little person who just arrived. Positive births help to promote your attachment and bond with your baby. A calm birth will also support you in being completely present to every precious moment of your newborn's life and will provide you with the emotional space that all new parents need to navigate your way through this life transition.

The flip side of the above scenario is the negative or stressful birth experience. Unfortunately, I hear about this all too frequently with second-time parents who find The Calm Birth School and those whom I previously worked with in my private hypnobirthing practice. Research has shown negative experiences during childbirth can lead to difficulties with bonding (Ayres, Eagle and Waring 2006, Nicholls & Ayres 2007) and with breastfeeding (Beck 2011). I have seen the hurt, anger and trauma from women who, when they look back at their birth experience, feel that they were not listened to or were ignored or disempowered. This is often coupled with feelings of isolation and the guilt of not being able to express the disappointment of their experience, for fear of appearing ungrateful for the healthy baby they hold in their arms. I hear stories of stress and angst from both mums and dads as they try to make their way through the negative rollercoaster of emotions that a stressful birth experience can trigger.

The message of this book is simple: *birth doesn't have to be like that.*

In this book, you will learn:

- Simple but powerful tools and techniques to help you feel calm, confident and at ease during your labour.

- How the mind-body connection can affect your ability to give birth optimally, and what you can do during your pregnancy to get it working *for* you, rather than against you.

- How to embrace your pregnant self and why it is important to do so for your birth.
- How you can get your medical care providers working with you for your specific needs, so that you feel like you are the one running the show during your pregnancy and birth.

- Some great ways to get your birth partner involved in your pregnancy, so they feel equipped to support you and be your advocate on the day.

I can't over emphasise how important your birth partner is going to be during this process, so if you know it's going to be an uphill challenge getting them to read this book with you, I highly recommend The Calm Birth School video course. This has all of the content split up into 10-20 minute videos that you can watch together in the comfort of your home. It feels even more like you're getting a private class from me. This, alongside the personalised support in the student's Facebook group, means you really can't go wrong.

Check out all the details for the video program here: www.thecalmbirthschool.com/course

Whether you read this book alongside studying the course or not, you will finish this book feeling uber-prepared for the most important day of your life.

Ready to get started?

Let's do this!

Suzy xo
www.thecalmbirthschool.com

Chapter 1:
How to Use This Book

Maybe you have gone mad and have a huge pile of pregnancy and birth books on your bed side table, or maybe this is the only book you intend to read before your baby arrives. (There are a million other things you could be doing, right? Because reading lots of pregnancy and birth books makes everything a bit too real!) Either way, this is the most important book you will read about creating a positive birth experience.

Within the pages of this book, and the amazing <u>Calm Birth School community</u> on Facebook, you'll find not only the tools you'll need for an amazing birth, but a philosophy to take into your life. You'll gain a better understanding of what you can control and what you can't, will develop an enhanced appreciation of the difference between flow and resistance, and you will understand what you need to do to adjust to your environment to suit you and your baby. Starting today and whilst reading this book, you'll gain a newfound inner confidence and will experience your unique ability to tune into the changes within your body as it grows with your baby. This will make it even clearer to understand exactly what you and your family will require to create the best possible foundation for the most positive birth experience that you deserve and desire.

To get the most out of this book, read each section and as you go along, and *complete the exercises and homework*. While it might be tempting to skim read, solely looking for the techniques for labour, The Calm Birth School takes a holistic approach to learning, so if you only look for bits about "how to breathe," you won't do yourself or your birth justice. I wish it could be as easy as saying read and breathe and you'll be fine, but it doesn't work like that. I know these techniques work, not just because I have used them during my own births, but because thousands of women have successfully used this exact same process too.

The line in the sand is clear when it comes to understanding the women who benefit most from these tools. The secret is, it's not really a secret. The success stories come from the women who put time into completing the homework and tune into the belief that they can create a positive birth experience, regardless of whether their baby decides to follow their birth preferences to the letter, or has other ideas.

The clients who asked questions, invested in <u>The Calm Birth School</u> course, bought books, did research and in some cases, got extra support when they needed it, focused on and strengthened their ability to create a positive experience.

Please note: at no point will I say that *positive* equates to *natural*. Your birth doesn't have to be a quick, natural, textbook birth for you to have an amazing experience. As you'll see in the stories of Calm Birth School clients throughout this book, each woman's birth was different,

including C-sections, long births, hospital births, home births and in some cases, like Laura's, very short births.

Laura's Story

Hazel Celia Ellerby was born on Tuesday evening, weighing 7lbs 1oz and after an extremely short but natural labour of only two and a half hours.

Here's my story: I was 40 +6 weeks and had a sweep booked for 3.30pm. Feeling positive and calm about the idea that this might slowly kickstart the journey to meeting my baby, I was quite surprised to hear that I was 2cm dilated, with a softening cervix and baby's head very low. The midwife hinted that I wouldn't need inducing the following week; she'd be out soon!

I lay on the bed and tried to concentrate on breathing while mum called the hospital. I think I even put on my affirmations MP3. The midwife asked to speak to me directly and advised that as my waters were clear I should progress at home. Being my first baby, I really wasn't sure what else to expect of these surges — if these were even surges, as my belly didn't move at all, it was all much lower down — but the midwife could hear how frequently I was zoning out of the conversation and my mum explained that all 4 of her kids were born very quickly. They advised us to go in for a check-up and to make sure my waters had indeed broken.

Mum called Julian who left work to meet us at the hospital, and then a taxi. By the time I'd made it down our tiny staircase, I'd had one surge at the top of the stairs and another at the bottom about 30 seconds later. I felt the cramps move to my bum and the need to push. Change of plan — get an ambulance!

The good news is that a paramedic arrived quite quickly, followed by an ambulance. They said I had time to get to the hospital so I was bundled into the ambulance and learned the beauty of gas and air. It really helped during the surges as well as over the 15 minute very bumpy journey. Coincidentally Julian met us in the lift on the way up to the ward (not that I was very aware at the time); he'd been asked to keep the lift open by the paramedics as they approached, and when he looked down at the wheelie bed he saw me! I went into a labour room, an out of action birthing pool in the corner, and asked whether a MLS water birth room was available instead. All my ideas of using the water, aromatherapy, bean bag and ball went out of the window when the midwife examined me: I was fully dilated, with baby's head in the perfect position for me to start bearing down during my next surge. Labour definitely didn't slow down during the journey to the hospital, then! I think I was so relieved that my mum wasn't going to be my midwife that my body let go!

The last half an hour I was taken off the gas and air to breathe as much oxygen down to the baby as possible, and was advised to make the most of each surge to push her out. For some reason, I was on my back, so swapped to leaning over the back of the bed on my knees, which immediately felt better. Julian was right beside me stroking my back, playing soft music on his phone and spraying lavender on the bed so it didn't smell too much like a hospital. The lights were dimmed too. My mum was about to leave us to it but she'd been such a big part of the action I asked her to stay. I had a lovely midwife and encouraging trainee so felt very safe.

I can't say I breathed baby out but the low humming I did whilst holding my breath and envisaging her coming out really helped. I reached down when I felt the head emerge and one last push brought her out and up into my arms. I held my baby in pure wonder. We had worked together as a team; I had trusted that she knew what to do, and the All is calm, all is well... I am safe mantra played in my head all throughout labour. Each surge brought her closer to me, but blimey, in the grand scale of things, there really weren't that many!

Thanks to The Calm Birth School, I had the natural, positive experience I had hoped for. And, despite it being super intense, it was a real bonus that it was so quick!

So how are you going to prepare for your birth? If you want the kind of birth that is available for you, you'll make a commitment now to do your homework and participate in the online group.

As you go through the book, highlight all of the parts that stand out as important to you and those that will be useful for your partner to read. I'm not going to lie to you, getting your partner to read through the relevant parts might be a bit like pulling teeth, which is why many of my clients find my videos helpful: it's easier to get them to sit down and listen rather than to read. Either way, don't skip getting your birth partner involved.

Even though you're the one having the baby, your birth partner plays an integral role in ensuring everything and everyone is set up optimally. Additionally, having a clear understanding of the techniques, of what to use, and of when and how to gently encourage and support you – without the worry of you morphing into Frankenstein's bride. We've all heard stories of women in labour who threaten physical violence to anyone within arm's length. Your partner's preparation is vitally important and is what will enable them to be the best support for you, alongside providing them with their own set of tools to fall back on. Witnessing this miraculous event is no small thing and when emotions are running high, 'your' hypnobirthing tools can be equally as invaluable to your partner as they are to you.

If you are enjoying a low risk pregnancy, hypnobirthing will help you to create a firm foundation for enjoying a natural, vaginal birth, with no intervention although of course this is not guaranteed. This is why reading all the entire book is important: I want you to be at ease with the idea you can navigate *any* situation that you are presented with in a way that leaves you feeling positive and confident that you are in control of your birth, rather than birth being something that *happens to you*.

A couple of vital hints: check out The Calm Birth School MP3s, listening to these daily will turbo charge your learning and your confidence. In addition, download your free Practice Schedule which you can access at http://www.thecalmbirthschool.com/bookbonuses.

Once you get into your final four weeks of pregnancy, re-read the book, step up your breathing and visualisation practice and you will be good to go.

So as we prepare to dive in, remember:

"Positive does not have to mean perfect."

- The Calm Birth School

Chapter 2:
What is Hypnobirthing?

When addressing the question of what hypnobirthing is, it's probably best to start with what hypnobirthing isn't.

Hypnobirthing is not hocus-pocus. As a former hypnotherapist and psychotherapist I laugh in the face of the idea of "mind control"... seriously, that is *not* what hypnobirthing is. The good news for you is that you won't find yourself barking uncontrollably every time you see a full moon or dry humping your birth partner in a room full of strangers for comic entertainment. Hypnobirthing has nothing to do with voodoo and *everything* to do with science.

So if it's not mind control or voodoo, what is it then? Quite simply, hypnobirthing The Calm Birth School way enables you to feel calm and connected to your baby during your pregnancy, providing you with invaluable tools and techniques that allow you stay relaxed during your birth. By the time you have completed this book - if you practise what I am going to teach you - you will be the equivalent of a Jedi Master at instant relaxation in whatever situation you find yourself in, as well as using your breath to feel calm and at ease, and utilising visualisation to help increase your focus.

Alongside the tangible tools are the equally important yet more intangible elements that impact your birthing journey. These present themselves in different ways for different people, but the common thread that underpins the change that many Calm Birth School students outline is a newfound confidence, self-belief in the birthing process, faith in your body, and not being afraid to share those beliefs with others. The Calm Birth School's approach to hypnobirthing aims to create the most positive birth experience for you, your baby and everyone involved in your birth.

Lisa's Story

My twin hypnobirthing story is bloody amazing!

Being pregnant with non-identical twins was a major shock: we discovered the girls at my 12 week scan and I will never forget the expletives that came out of my mouth. I think I cried for a week.

It soon became clear to me that I would have to relinquish some control to a specialist midwife (and when I say specialist I mean useless) and go from my lovely idea of giving birth in my local birthing centre in the water to a possible induction or even C-section in a hospital. I was told this was standard protocol for twins, even low risk twins like mine.

This wasn't how I'd imagined it! I was fed horror stories about what could happen to my babies and how small they would be. I was laughed at by a doctor when I requested a water birth, and another doctor even told me to take an epidural because, and I quote, "Even if twin two is in the right position, I may have put my hands inside you and pull it down the birth canal."

From that moment onwards I vowed I would take back control and that I would have the peaceful water birth that I and my babies deserved.

I contacted Suzy from The Calm Birth School and started The Calm Birth course. Suzy and I also had a one-to-one Skype call which was very encouraging and gave me the confidence to contact the midwife supervisor and complain about the way I was being treated. Thankfully she completely agreed with me and we put a plan into place. I had to plan for the "what ifs" but I didn't write these down; I only wrote down the birth I wanted.

I put my consultant's nose out of joint by refusing an induction between 36-38 weeks. I felt confident that my body would know what to do when the time was right. I had saved The Calm Birth School's affirmations to my phone and I would practice the breathing exercises and do my birth rehearsal almost nightly in the bath, as well as going over the videos daily and listening to Bob Marley. I swear that Bob got me through the last few months of pregnancy.

I was given two sweeps a week apart from each other, both of which made me surge. The surges however came to nothing. I got to 39 +2. This was no easy feat! The surges had started at 37 weeks and sometimes they came every two minutes and were very strong, so strong that I even took myself to hospital, only for them to then stop.

At week 39, I reluctantly went along to my consultant's appointment where I gave in and let her book me for an induction.

Initially I had been determined however I wasn't going to go along, but the previous couple of weeks and all the surges I'd experienced had taken their toll on me. I was wound up and anxious and I felt like I was pushing my kids, who were getting so over-excited to meet their sisters, from pillar to post.

I must admit I got home from that appointment and cried. I felt like my body had let me down and I couldn't believe that something I'd fought against so strongly was going to happen.

We arrived at the hospital for induction the next day. The midwife looking after me gave me a very good sweep before popping the Propess inside me. She left the room and told me it could take 24 hours even for ladies who have birthed before, and to have a walk and get some lunch. Within ten minutes I was looking at my husband and telling him that the surges were strong. Looking at the monitor I was strapped to, I could see that they were peaking.

We called the midwife back where she checked me and said I was 4cm and my waters were bulging. She pulled out the Propess and said, "Well that's not needed then!" I was so relieved. Forty minutes later, I was begging to get into the birthing pool.

Once I entered the water I felt all the build up of tension leave my body. I used my TCBS breathing techniques and my husband played Bob Marley for me on repeat . I used the breathing techniques for four hours before I started to feel the urge to bear down. At that point I grabbed the gas and air to birth Indy.

There was suddenly an influx of doctors dressed in scrubs with a scanning machine asking me to get out the water to scan me. I refused. They began to scan me in the water and the doctor scanning me said, "I can't see a head," to which I replied, "That's because I'm pushing it out." Indy arrived, and we welcomed Zadie five minutes after her sister. Indy Florence weighed 7lb and Zadie George weighed 6lb.

Even though I doubted myself, I managed to give the perfect birth to my girls. Everything just fell into place, even down to the midwife who was on bank staff at the hospital from a local birthing centre where they champion natural water births. She was so encouraging. I had my placentas encapsulated and the girls latched on to feed within half an hour of being born.

I would champion hypnobirthing all day every day! It really helped me to draw on my inner warrior when I needed her. Now, I'm even thinking of training as a doula, specialising in twin births to show my fellow twin mamas that we don't have to be scare-mongered.

Thank you TCBS (and thank you Instagram for helping me to find it).

How Do You Create Your Own Calm Birth?

It's rare that women see other women giving birth in Western society today, apart from the overly dramatised births depicted on TV shows. Labour is a secretive event that takes place in a hidden away ward in a hospital or behind the closed doors of your own home. However, this huge shift towards privacy has really only come about in the last forty-five years or so. Nevertheless, it has made an unimaginably massive difference in the way we view and experience birth. Many women will have never been around other women giving birth until they are in labour themselves.

This has left us in an interesting place. For many people, the only insight we get into one of the most significant rites of passage we will experience comes from TV and the movies, where it's often sensationalised, dramatised and presented as a torturous experience: a woman lies on her back, her face screwed up in agony, and she screams and pushes and writhes in pain until her baby comes out. This has skewed our perception of what labour is like. We go into birth fearing the worst because, let's be honest, nine times out of ten, what we have seen in the media looks bloody horrendous. Our sanitised culture has also invited us to be squeamish and recoil from anything involving bodily fluids.

On rare occasions, we remember that we're actually mammals and might look to how other warm-blooded animals give birth, via a documentary for example. It's easy to see an animal's four legs compared to our two and to assume that birth is going to be easier, quicker and less painful for animals because "we humans are just built differently."

The problem with this assumption is that it's just plain wrong. Our physical design is not the issue. If that was the case, how do you explain the thousands of women who give birth every year using hypnobirthing techniques — often without any pain relief — who recall their births as easy, comfortable and joyful? You may be tempted to write off those experiences because these women have higher pain thresholds, bigger pelvises, gave birth to smaller babies, or any of the other excuses, I mean stories, that spring to mind. These are however not the real reasons why some women's birth experiences are so empowering. There is more at work here.

While the wide scale medicalisation of birth means we are fortunate to be able to tap into world-class medical expertise in the West should we need it, it also means more women opt to birth in hospitals rather than their homes. Where women once witnessed other women giving birth many times as mothers, sisters and friends, therefore thinking of birth as normal and natural, nowadays birth is often perceived as a mysterious, scary and necessary evil, one we have to endure to get to the 'good bit' at the end. This has left us feeling less confident about our ability to give birth – a job we were made for – without medical intervention and assistance.

There is a simple reason why a woman who doesn't do the type of preparation you are about to undertake will be unlikely to give birth in the same way as any other female mammal on the planet — comfortably, calmly and without fear. This reason has nothing to do with design features. In fact, for normal, straightforward pregnancies, the pleasant delivery of a newborn has *everything* to do with what's going on in your mind as you prepare for and give birth. The real issue is in the mind, not the body. Now that we know this, we can address it.

The Evolution of Birth

When we look at evolution, female physiology, and the fact that women have ensured the survival of the human race during the last two hundred thousand years, it's fair to say we have done a pretty good job, right? Our ancestors somehow instinctively knew how to give birth. They would have been exposed to it during everyday life as I mentioned above, which demystified it and stopped it from being scary. This genetic wisdom has been imprinted in your DNA. I see it as The Calm Birth School's job to provide you with the tools and techniques you need to access that information. This is knowledge that you have always had at your fingertips but which has been drowned out by the cultural noise we live with in Western society in recent decades. Luckily, there's no rocket science involved in getting tuned in, but we do need to get a bit scientific.

Biology and Neuroscience

I'm not a scientist. However, the fundamentals of hypnobirthing and how it works are based on the biology of the body and the science of the brain, otherwise known as neuroscience.

For the vast majority of women, the instinctive response to the onset of labour is physical and emotional stress. This stress response is triggered in the limbic system, which sits in the oldest part of the brain, formed around 200,000-250,000 years ago.
When we look back to those times, the most important aspects of human life were reproduction and survival (some argue that they still are). Language did not play a role in life; everything was based on the messages we received via our five senses of sight, touch, sound, taste and scent. These cues

would trigger chemical and physiological changes within the body, determining how we felt and what we needed to do, whether that meant running for our lives, fighting to the death or freezing and pretending we were dead in order to protect ourselves.

The part of our brains responsible for logical, rational thinking and critical analysis, the neocortex, didn't even exist at this point. Life centred around the instinctive responses needed for survival, in order to protect ourselves from the dangers and threats of the outside world.

This ancient hardwiring is present in us even now. The limbic system continues to protect us from danger in exactly the same way today as it did hundreds of thousands of years ago. When we receive a danger cue (whether real or imaginary) the limbic system, which was designed to keep us safe, wins in the fight between instinct and thinking about the best course of action logically. In short, it overrides the neocortex and tips us into fight-flight-freeze mode.

When we bring this back to the context of birth, although most women in the West are not facing life or death scenarios, the fear that arises when giving birth inevitably triggers our fight-flight-freeze response. In turn, this sets off a physiological and emotional chain of events which prevent us from birthing optimally in the way we were designed to do.

Part of The Calm Birth School's approach to hypnobirthing is ensuring that you understand how you can minimise the fight, flight or freeze response by working *with* the biology of the body, instead of against it.

In addition, this approach will help you to start thinking and feeling differently around the expectations you have about birth. This will support you before, during and after the actual act of giving birth.

The Neuroscience Bit

I'm guessing you've heard the about a technique called visualisation. This is another one of those words that people can think is a bit hokey. However, it is to become a vital tool in your hypnobirthing journey and you'll find masses of information, evidence and case studies on it when you look at psychology and the science of sport.

Before Tiger Woods became known as 'the adulterer,' he was most widely acclaimed as the greatest golfer the world had ever seen — and he is an avid visualiser. You might also have heard of a guy named Jim Carrey. What about Will Smith? David Beckham? Oprah? *Come on now, you must have heard of Oprah!* Seriously though, all of these multimillion-dollar high achievers have spoken widely on the power of visualisation. So what has this got to do with how you can give birth calmly, comfortably and more positively? It's all about your wiring.

Whether you actually see something in real life or simply imagine it, the combination of what you perceive and how you feel about what you are perceiving creates what is known as a *neural pathway* in your brain. This pathway helps you determine how you are going to respond to stimuli that come your way. Think of it as a blueprint that your brain calls upon when you are faced with a situation where you feel threatened.

When it comes to birth, whether you have consciously thought about your forthcoming experience or not, during your life you will have seen many negative images and depictions of what birth is going to be like, as we discussed above. Even before pregnancy, many of you have already created a less-than-wonderful birthing blueprint, which whether based on fact or fiction, draw a woman's body into fight-flight-freeze mode instinctively, because it triggers the suggestion that there is danger, a need for fear or concern about your ability to birth effectively and efficiently. This gets overridden if you have witnessed or been a part of a positive birth experience, when a new blueprint can be formed through exposure to new stimuli.

David Beckham, Tiger Woods and the like use visualisation to create the ideal blueprint of their role within the game or in the tournament before it has actually happened. They teach themselves how to kick or hit the ball at the perfect trajectory in their minds, many, many times. Then, when the time comes to step out onto the pitch or golf course, their brains instinctively know where to go because their present situation feels so familiar. This is not a new situation to them; they have in fact already visited the scene many times through the practice of visualisation. And because they are familiar with it, they do not have to deal with the same levels of adrenaline and the stress hormone, cortisol, that come from being plonked into an unfamiliar scenario. The fight-flight-freeze response doesn't kick in, which in turn allows them to perform at the peak of their potential.

This stuff works just as effectively in a labour room as it does on the football pitch, and I am going to show you how over the next few chapters.

So let's look at what you are going to learn and take action on:

1. How to manage the instinctive fight-flight-freeze response.
2. How to minimise your physiological and emotional responses to stress.
3. How to promote oxytocin and endorphin production to aid the process of birth.
4. How to create the optimal environment for giving birth.
5. How to ensure your birth partner is able to work confidently with you and your care providers.

The key element that ties everything together is learning how to *relax*. Sounds good, right? It is. Let's keep going; there's a lot to cover!

Chapter 3: An Introduction to Relaxation

Reasons to Relax

Mindfulness and meditation are buzz words nowadays, and they have entered the mainstream in a major way. Global organisations such as Google, PricewaterhouseCooper and even the UK Home Office have embraced mindfulness training for their employees. In October 2015, the UK government pledged to invest £1million per year into training school teachers about how to teach mindfulness to children. As we catch glimpses of the latest articles when scanning social media or trade and fashion magazines, we know there's something in this consistent drip-drip message of switching off, but many of us just can't bring ourselves to do it.

The fear of missing out on what's going on in cyber space often robs us of the invaluable time we need by ourselves. Believe it or not, *you* need to come at the top of the priority list during pregnancy, followed by your loved ones, family and friends, with the social media world at the very bottom of the list.

The first thing I invite – no – *challenge* you to do during your pregnancy is to start deliberately taking time for yourself. In an ideal world, you would enjoy some quiet time for thirty minutes at least once a day. That means no television, no phone, no laptop, not even a book: just time and space where you give yourself permission to let your mind have a break from thinking, analysing, planning and doing. Some of you will read that and think, "Hell no! That sounds hideous!" Others will think, "Yes that sounds great, but back in the real world…" I get it. You're busy. However, you're also pregnant. You're growing new life inside you. If thirty minutes feels overwhelming or unrealistic for your life right now, start with less.

For example, you could do fifteen minutes of nothing before you get out of bed, focusing only on your breath using one of the breathing techniques you're about to learn. If even that's not possible, break it down into five minutes chunks. Get unplugged three times throughout the day and focus on nothing but regulating your breathing (go and lock yourself in the toilet if you have to - I promise no one will disturb you in there). Should things pop into your mind, acknowledge them, tell them you'll address them later, and bring your attention back to your breath.

There are lots of reasons why this is good for you and your baby, so don't resist it. You're reading this book to give yourself the best possible foundation for creating a positive birth experience for you and your little bundle of fun, and the preparation starts here, oddly, with choosing to do nothing.

It's a pretty good gig and if I haven't hammered home the message enough, here's a list of the reasons you've got to get with me on this one.

1. It will lower your stress levels – this is a BIG deal
A lot of women work or live with more than their fair share of stress. When you take the time to switch off, it actively lowers the amount of the stress hormone, cortisol, in your blood. You'll appreciate this in ways you might not even notice, but trust me: low cortisol levels are a good thing. Your baby agrees too.

2. It helps you think more effectively
When we're stressed or angry, we become stupid. I'm being blunt but fair here. We find it difficult to think rationally or creatively when we're in a stressed out state and are essentially useless at problem solving. That's why The Calm Birth School breathing technique is especially useful when you're in a stressful situation at work (we'll outline the technique a little later in the chapter). Ideally you would excuse yourself from the problem, go to the loo and do a bit of TCBS breathing, in order to create the mental space to come up with a solution. Even if you can't excuse yourself, the simple act of remaining silent and focusing on your breath will help you disengage from the frustration and be more proactive rather than reactive.

3. It helps you control pain
There is more and more evidence to suggest that the more we are able to disengage with our surroundings, such as when we

go to that meditative state of mind, our pain receptors are less sensitive (Ziedan 2011).

4. It makes you more emotionally calm
Not only are you able to handle stressful situations more easily, but you don't get riled as often either. Stressful situations roll off of you like water off a duck's back, which is good for you, good for your family and amazing for your baby.

The feedback I have received about the benefits of these tools during my clients' hypnobirthing journeys is that these techniques are not just indispensable during birth; they also become useful tools to draw upon after baby arrives. Being able to stay calm when you feel completely responsible for this new little life whilst simultaneously being completely out of your comfort zone is an invaluable life skill. Start putting yourself first today and embracing the downtime that is oh so good for you.

Hypnosis/Trance

In many cases, people feel weird about hypnosis (which is often referred to as trance, and I will use the two terms interchangeably). The sense of weirdness often comes from what they have seen on stage and TV shows. While this type of hypnosis often makes for great entertainment, it is completely misleading and incredibly frustrating for anyone who uses hypnosis in a therapeutic setting because what we see on TV or on stage is an illusion: the illusion of mind control.

Let me explain what I mean. The first thing any person who works in the field of hypnosis is taught is that all hypnosis is self-hypnosis. This means that you can't make anyone do anything that doesn't sit comfortably with them. When we watch a stage hypnotist, what we're seeing is an audience made up of two kinds of people: those who are there to be entertained and those who are there to be the entertainment. We'll see a large group of people invited up onto the stage and they'll be directed to participate in 'entertaining' things that usually involve them looking a bit silly. As the show goes on, the suggestions that the stage hypnotist uses will become increasingly outrageous. Every single person up there will have their own internal barometer that tells them when enough is enough. As each person reaches that point, you will notice them choosing not to participate on the stage, and at that point the hypnotist will tap them on the shoulder and ask them to sit down. Finally, you are left with the 'star' of the show — the person who feels the least inhibited in that environment, doing things that under normal circumstances you wouldn't want your grandmother to watch. At no point can anyone be forced to do anything they do not want to do, and the only reason that they are clucking like a chicken or barking like a dog on demand is *because they want to.*

I hope that sets the record straight once and for all: hypnosis has nothing to do with mind control!

So if *isn't* mind control, what is it? Hypnosis is a natural state of consciousness we drift in and out of whenever we have a narrowed focus of attention or are deeply relaxed in passive or active activity. It's something that we all drift in and out of naturally every single day at least ten to twelve times. The two times that you will be most familiar with are when you are just drifting off to sleep or when you are just waking up.

Other examples of natural states of trance include while:

Daydreaming
Running
Sex
Drawing or colouring in
Driving
Watching a great film
Listening to someone tell a story

We enter into trance whenever we experience a narrowed focus of attention, becoming deeply absorbed in an activity that we are very familiar with or are extremely relaxed doing, the logical part of our mind becomes more easily sidetracked. Whenever the brain is not required to critically analyse everything we are doing, a gap in our thinking is created, and this leaves us more open to suggestion. This state of suggestibility cannot occur when we are learning a new skill or taking part in an activity that requires us to be alert and aware.

This is really important to remember if you find yourself listening intently to someone else's birth story, particularly if that story is about a terrible experience. The more involved and engaged you become with the story, the more open to suggestion you become. In this type of scenario your brain stops critically acknowledging that the story you're hearing is only the specific storyteller's experience, and without realising it you start to file the information under the 'What Is True About Birth' file in the brain's hard drive. As the suggestion that birth is an awful and scary experience gets implanted in your brain through listening to this tale, it becomes your point of reference when you think about birth and most importantly when you go into labour.

Therefore, it is imperative that you consciously remove yourself from listening to negative birth stories at any opportunity. However, it's not all bad. The fact that you're more open to suggestion and to storing information in your brain's database when you feel calm, relaxed or focused is *exactly* what we capitalise on when it comes to your hypnobirthing journey. This is the exact reason why hypnosis is so great for helping you to retrain your mind about how to think and feel about birth. It can be used just as effectively to reinforce positive suggestions, feelings and associations you have about your upcoming labour and birth.

So let's bust some hypnosis myths:

You can get stuck in hypnosis
You can't get stuck in hypnosis in the same way that it's impossible to get stuck asleep!

Hypnosis is harmful

The only time hypnosis can be harmful is when a person is in a natural state of trance, listening to someone else's suggestions about an event uncritically, such as in the example given above, when you finds yourself listening to someone else's negative birth story without acknowledging that the experience being recounted does *not* have to have any bearing to the way that you give birth.

Hypnosis is mind control
Well, we've already busted that one! All hypnosis is self-hypnosis, remember?

The Brain

While we're not going to dive deep into the truly fascinating world of neuroscience, understanding the basics of way the brain works will help you get clear on why going into a state of trance, self-hypnosis or meditation by yourself is so great for birth.

I'd like you to imagine the brain being like an iceberg. The tip of the iceberg is where all of our logical, rational thinking takes place, but the subconscious, which is effectively all the stuff going on beneath the surface, is actually where most of the action is.

The subconscious mind runs our autonomic nervous system, and is responsible for the beating of our hearts, the blinking of our eyes and anything we do without consciously thinking about it. It is also the home of our emotions, our imagination and where our gut feeling or 'sixth sense' originates.

The driving force behind everything our subconscious does is ultimately designed to keep us safe. The way it does this is by keeping records or memories – both visually and emotionally – of everything we have ever seen or been a part of, and categorising them crudely as either 'safe' and 'unsafe.' If we have no frame of reference for the event taking place, the subconscious will look for the most similar thing we have experienced and will trigger a set of physical and emotional responses based on that event.

An example of this might be someone witnessing another person get hit by a car. If they haven't been able to logically process the incident, the memory of seeing the accident will trigger a chain of physiological and emotional responses, which might result in the subconscious telling the person instinctively, "It's no longer safe to cross the road."

This is what happens so frequently when it comes to birth. Unless you have experienced a difficult birth, the majority of women will have only witnessed challenging births on TV or heard other people's stories. As you listen to or watch these stories and events, you experience a negative charge of emotion, and the brain files the situation or the image into the 'unsafe' database. Should you ever be faced with a similar situation (i.e. if you go into labour yourself), the subconscious mind will try to protect you from having your baby by releasing hormones that activate your survival mode – the fight, flight or freeze response that we discussed in chapter one. This impedes your ability to birth calmly, comfortably and efficiently as your subconscious attempts to protect you from this 'unsafe' event.

The part of the brain responsible for activating the fight, flight or freeze mechanism sits in the oldest part of the human brain, the amygdala. The amygdala doesn't understand language or logic. It responds to situations based on imagery and feeling: what we see, even if only in our imaginations, and how we feel at a particular time.

It's extremely interesting to me that the hormonal and physical responses of the body are incredibly similar whether a person is seeing an event for real or imagining or dreaming about a situation. This is why, if you're having a nightmare, your skin feels damp from sweat and you experience an increase in your heart rate. Even though you are dreaming, the brain cannot tell the difference between what is real and what is imagined. It responds to what you're seeing and feeling in your dream in the same way as it would if you were having the experience in real life. Let me reiterate: *the brain cannot distinguish what is real from what is unreal.* If you perceive an event to be 'true,' either through your physical eyes or purely through your imagination, the brain will process that event as if it were actually happening, and your body will respond accordingly.

In order to change the physiological chain of events that occur when you have either consciously or subconsciously decided that birth is unsafe, your task is to literally retrain your brain to expect something different.

This is achieved by imagining your ideal birth scenario and regularly listening to new audio input such as MP3 meditations. This simple process is so powerful because it enables you to begin to create a whole database of 'memories' about how positive your birth is going to be, meaning you can change which category your 'birth' files are stored in, moving them from being labelled as unsafe and scary to being safe and positive. You do this by retraining the brain to expect something different in this type of situation.

Repeated exposure to new audio input, combined with watching hypnobirthing mums give birth (yes, I recommend you doing that too), will begin to build up a positive mental picture of how your birth can be. As you picture your birth more and more frequently, imagining how it will feel at the various stages, adding emotion to your future birth story, new files will be added to your subconscious database to cross reference with when you go into labour.

It's funny: as you practice relaxing, switching off and doing nothing every day throughout your pregnancy, you will actually be achieving so much! The skill of being able to relax very, very deeply on demand will become both instinctive and invaluable by the time you reach your labour day. You will look forward to each surge, using it as a sign that you are closer to meeting your baby. You will be able to relax and enjoy (yes, I said *enjoy*) the process, as you learn to trust that your body and baby know what to do at the deepest level.

The Calm Birth School Breathing Technique

The first technique you will learn is TCBS breathing. This is a simple, powerful technique. The reason it is so effective is because it triggers the body's natural calming reflex, which occurs when the out breath is nearly twice as long as the in breath. During your labour, this technique will help you maintain a deep state of calm and you'll use it in between surges. You will also use it as you feel a surge coming in and once it has subsided.

Ideally, the breath is taken in and out through the nose as opposed to the mouth as this gives you more control over the flow of air. However, please don't stress if you have a cold when you're birthing. Just breathe through your mouth – it will all be okay.

How to do it
Breathe in deeply to the count of four through the nose, imagining you are filling your lungs right to the bottom. As you breathe out, imagine sending the breath down, so it moves around your baby, down your legs into the tips of your toes and then into the floor. Sometimes when you are first practicing this technique, it's useful to put your hands on either side of your waist, so you can feel the rise and fall as you breathe deeply. It really is as simple as that.

When to use it
Use this technique whenever you find yourself feeling stressed, whether it be at work, with your partner, getting on

and off public transport - wherever and whenever. This will help you to relax quickly the more that you practice it. In addition, should you be one of the many of ladies who labour quickly, this technique will get you into the birthing zone quickly and easily once you have become accustomed to using it in your everyday life.

If you are the type of person who tends to take life in her stride with very little stress, this doesn't make you exempt from practicing. It's just as important you carve out some practice time for your TCBS breathing too. Ensure you do one set of Calm Birth School breathing in the morning for five minutes, five minutes at lunch time and five minutes again in the evening.

Emiliana's Story

After a manic few weeks of moving house on the 28th August (my guess date being the Friday 4th September), I decided to treat myself to a pregnancy massage on my guess date. We had told bump that the 5th September would be the optimal date she could come: baby's father James would be home from work and it would give us enough time to settle in the new house (not unpack, just settle!). The massage was the best feeling ever and included some visualisation of baby arriving and the use of clary sage oil which was so calming and lovely. James said I looked like a new person when I got home as I looked so relaxed and healthy. We had friends over for an easy pizza dinner that evening and I'm sure the oxytocin from the massage and fun with friends got things moving...

I didn't feel any different when I went to bed at 1am that night, but the surges started at 6am that morning, feeling like mild period pains. I knew they were surges because when I timed them they were evenly spaced apart.

I woke James up at 6:45am and he kicked into action, making the living room all lovely with candles and incense. I ate some cereal and drank tea, thinking of TCBS stories I'd read where things still took a little while from here. As we were hoping for a home birth, we decided to call the midwife at around 8am. She said she'd be at least another half an hour so I carried on welcoming the surges and laying out the collection of birthing tools (jelly babies, water, straws, paracetamol and so on) that I might need, as well as getting the TENS machine going.

At 8:30am the midwife came and I was 3cm dilated. I felt a bit disappointed and immediately Suzy's voice popped in my head saying "It's only a snapshot in time!" She was totally right: an hour and a half later, after a bath and more powerful surges, I was 7cm.

After some funky standing up rocking movements with James holding onto me from behind (we are still laughing at how ridiculous we must have looked but it seemed like the only way to ease the discomfort!), my waters released. I got into the pool just before 11am. That's when the midwife noted active labour starting. The second midwife then arrived and the show really got on the road!

I took to a few different positions in the pool over the next three hours: head against the side taking in gas and air, and then back around, sat on the inflatable stool with legs up and my hands grasping the back of my legs. James was always right behind me whispering words of encouragement and handing me the gas and air.

During the last half an hour, the midwife respected my TCBS notes (she did the whole time but I particularly noticed it here), and talked about breathing the baby down and what it would feel like when she was finally in my arms. James did the same and I swear it was the visualisation and allowing my body to take over in those three hours that got me and Mollie through in the way we did. I didn't experience any tearing; again I put this down to letting my body do its thing rather than push at the wrong times.

However, by the time her head was out, I had lost all my energy. I heard the midwife talking about Mollie's heart rate creeping up which spurred me on to give the next surge a big helping hand. It didn't come for about three minutes and James said it was the longest three minutes of his life! But I talked to Mollie saying we had to do this one together as the team we were. She absolutely shot out when that final surge came, and the cord snapped because of a combination of this and it being shorter than usual.

Mollie's end clamped itself naturally (amazing!), but I lost 400ml of blood, which I didn't notice at all with the euphoria of holding our newborn baby. I took the injection to birth the placenta as it wasn't coming naturally and with the blood loss the midwives strongly advised it. By that point, I was so happy with the way everything had gone that I didn't mind some intervention to make it easier to spend time with our little baby as a new family. Active labour was recorded as 3 hours and 8 minutes. I put the whole experience down to the TCBS techniques, a positive mindset, being at home (I'm sure this was only possible because of the positive mindset is coached myself into) and the birthing pool.

So, a big thank you to Suzy from James, Mollie and I. It's been an amazing start to our journey that we know will influence how we are as parents and how Mollie will develop too. We hope we will be using the TCBS techniques again in the future!

Chapter 4: The Biology Bit

Whilst some sorely mistaken people are under the misapprehension that hypnobirthing is all a bit mumbo jumbo, you, alongside your birth partner, should be starting to understand that the whole foundation of hypnobirthing is based in science.

The Uterus and What Happens in Labour

No one actually knows why we go into labour precisely when we do. If I knew the formula to that one, I'd be a very rich lady indeed! What we *do* know is that the body generates a huge surge of the hormone oxytocin, which I will explain more about in just a little bit. The large quantity of oxytocin in the mother's system stimulates the muscles of the uterus and kickstarts the process of labour.

There are two types of muscle in the human body: voluntary, such as those in the arms and legs, and involuntary. We have conscious control over the movement of voluntary muscles, but we are unable to consciously control involuntary muscles.

The uterus is made up of two layers of involuntary muscles, one horizontal and one vertical, that work together as a pair.

They operate in the same way as any other muscle in the body, the only difference being, as I've just mentioned, that you cannot consciously move or control them. When you go into labour, the vertical layer of muscles moves down over the horizontal layer and begins to pull the horizontal layer upwards. When a woman is calm and relaxed, this movement is smooth, the muscles working together in harmony. The upward motion causes the neck of the cervix to thin out and open up so that baby can move down the birth path easily, without stress.

You can see what this looks like here:

Inner layer
The inner layer of the uterus is made up of circular horizontal muscles, which are located in the lower area of the uterus with the thickest ones situated just above the opening or neck of the uterus (the cervix). In order for the baby to move easily down into the birth path, these thicker muscles have to be drawn up and back.

Outer layer

The outer layer of the uterus is made up of stronger vertical muscles. These muscles go up the back and over the top of the uterus, drawing up the relaxed circular muscles of the inner layer.

The Uterus in Surge

When the birthing mother is in a state of relaxation, the two sets of muscles work in harmony in a wave-like motion. The vertical muscles draw up, flex and expel, and the inner circular muscles relax, open and draw back. Birthing then takes place smoothly and easily. Remember what I said in chapter two: your body is designed to give birth.

When you have a surge during labour, you may notice your tummy gets very tight and you can see things noticeably lifting upwards, as shown by the dashed line in the diagram. The uterus lifts up as the muscles work during a surge, and then return to their normal position.

When a woman is tense or scared, the involuntary muscles of the uterus are prevented from working together in harmony. As the horizontal muscles tense up, the vertical muscles attempt to pull up these rigid, inflexible horizontal muscles. This state of muscular tension, along with your baby's head putting pressure on a cervix that is not thinning or opening because of lack of movement, causes the pain that many women experience during labour. This experience confirms a woman's initial fear that labour is indeed painful, which keeps her in a tense, stressed state.

Veronique Strohbach, author of *What Size Is Your Brain?*, states that, "For every thought we have there is a corresponding physical and chemical reaction in the body." Assuming we have healthy mums and healthy babies, this quote becomes the cornerstone of our understanding around how and why the uterus either works efficiently or becomes tense and inefficient when we are birthing.

Earlier, I explained the brain's database of memories and how it sends messages to the body about what is happening. For some women, another person's story will flash through their mind during labour, sending out a red alert that something is wrong. For others, something or someone within the birthing environment may trigger feelings of unease or being unsafe, which in turn puts the body into survival mode.

One of the body's key chemical responses to a conscious or subconscious thought of danger or threat is the release of adrenaline. Adrenaline isn't a friend of labour. When we go into survival mode the physical response of the body is to divert blood and oxygen *away* from the uterus, because in evolutionary terms, the uterus would not have kept us safe from whatever the danger – real or perceived – was. As much blood and oxygen as possible is sent to our extremities because of their ability to help us fight for our lives (with our hands) or flee for our lives (with our feet). The uterus is understandably deemed useless and, as such, it is starved of the blood and oxygen it needs to function and becomes less and less efficient.

This is great if you're about to be attacked by a lion, but when it comes to labour, this is not a good scenario: the lack of blood and oxygen available not only slow the process down, but also make it more painful. The pain then reinforces the initial thoughts of fear, confirming there really is something to worry about. As labour slows and the birthing mother becomes tired, she is much more likely to need some kind of intervention as her body and baby become less able to cope with the impact of the stress response.

In other words: fear equals tension equals pain.

Whenever you are feeling stressed, but particularly in the run up to and during labour, one of the best things you can do to counteract fight or flight mode is to bring your attention to your breath, and focus. The Calm Birth School breathing will help to neutralise the stress response and create a release of endorphins.

Why Endorphins Are Your New Best Friend

Endorphins are amazing for birth and a great friend of labour. They act in a similar way to opiates in the body, and are said to be up to 200 times more effective than morphine when it comes to pain relief, which is why you ought to love them. The more things you can do to relax and feel good during the early stages of your labour, the more of those happy hormones you will bank, helping you feel more at ease, more in control and more comfortable.

The other hormone I mentioned earlier, and one you need for an efficient labour, is oxytocin. Oxytocin (pronounced oxy-toe-sin) is "the hormone of love," a chemical we release into our body whenever we feel love for something or someone. When you make love to your partner, your body courses with oxytocin. (It's the reason many of you are reading this book with your lovely bump!) Oxytocin is the hormone that gets labour started. When you are able to stay feeling calm and at ease, it's also the hormone that causes some women to feel euphoric and even joyful whilst birthing. A woman receives the biggest peak of oxytocin at the end of labour. This happens in between birthing baby and expelling the placenta (or afterbirth as some people like to refer to it).

Oxytocin is the reason why, no matter how moulded, squished or messy baby's head will appear when they are born, many women look down into their newborn's eyes and fall totally and utterly in love at first sight.

This is part of nature's very clever evolutionary tool. It makes us want to stay close and look after our babies, which ensures the survival of the human race. The physiology of birth remains pretty much the same as it was over 200,000 years ago when homo sapiens first appeared. Not only did nature get the third stage right, birthing the placenta, but it also got the first and second stage, early labour and the birthing phase, pretty spot on too. Healthy women needed to be able to give birth as efficiently and as enjoyably as possible, because when we re-visit what early human society looked like, there were no doctors, midwives or doulas (in the form that we know them, at least) at all. Yet women still gave birth again and again many times over, until we found ourselves here in the twenty-first century.

In short, *we women were designed to do this job*. As we release resistance in the lead up to our birth and deepen our trust in the knowledge that our bodies and babies know what to do, when we reach our labour day, we really can let go and follow whichever direction our baby decides to take us in.

Pain

The time has come to put something on the table: we need to talk about pain. By now you should have a very clear understanding of the science behind why some women birth really comfortably, whilst others do not. As I said a few moments ago there is a relationship between the kind of thoughts you think, the amount of fear you feel, how tense the body becomes as a result of that fear, and how much pain you experience. Once again:

Fear = Tension = Pain

Does this automatically mean that if you're completely fearless about your birth, you won't experience any pain? No: there are no hard and fast rules about how you will experience the sensations in your body when you are giving birth.

Some of you reading this book will birth in complete and utter comfort, feeling relaxed and open, without resistance, able to birth your baby whilst feeling *sensations,* but not *pain.* Others of you will feel sensations that you might have previously described as pain, but through preparing for your birth and mastering the techniques in this book, you will feel that the experience is totally manageable. You'll find that you are able to distance yourself from the experience, by using your Calm Birth School techniques to keep you in the zone. Or you might also get to the other side of your labour and say, "It was bloody painful, but I rocked it!" Every woman's experience is subjective and not something you can measure your own birth against. All of these kinds of births are positive hypnobirthing experiences.

The key marker of the women who do experience intense or painful births who also have the tools of hypnobirthing in their back pocket is that they understand that whatever they are experiencing is totally normal and nothing to be fearful of. This allows them to redirect their focus and choose to experience the physical sensations in a different way.

Pain is subjective and we all experience it differently, depending on how we are feeling emotionally and what our brain decides is the most important focus at that moment. If you've got bad period pain but have a lot to do, you can power through it, focusing on the task at hand rather than worrying about how your abdomen feels. But the minute you finish your chores and sit down, how bad you're feeling suddenly hits you. It's all about your focus.

If, when you're birthing, you think about what a nightmare everything is, when are you going to feel the next surge and for how long, the whole experience becomes tiring, draining and intense. Rather than focus on how painful labour is going to be, it will be much more helpful for you to be curious about which techniques are going to help you remain comfortable.

I also recommend that you explicitly ask anyone involved in your birth to talk about your *comfort level* as opposed to how much pain you're in — more on this in chapter 8 when we address communicating with your care provider. Remember that when you're sat in a state of relaxed consciousness, you will be far more open to suggestion; we don't want a well-meaning midwife making an indirect suggestion that you need assistance, and for you to start to analyse that and take those suggestions on board.

A woman who is able to look at the sensations within her body as part of the normal process of birth, feeling at ease with what she's experiencing and knowing she doesn't need to fight against it finds the whole process much easier. She becomes able to experience the intensity of birth without fear, knowing that the sensations she is feeling are normal and natural, and reminding herself that all she has to do is allow her body to go with it. She releases all resistance. The way she experiences that intensity or pain becomes a wave that she is able to ride, rather than something she is trying to fight against, as Karis' story below demonstrates.

Karis' Story

At week 41 +3 I relented and had an appointment for a sweep. After having refused the first one the week before, we made the decision, with the help of The Calm Birth School, that a sweep was really okay and may help things move on a bit. At this point I should say that my parents had come over from the UK for one of the guess dates and I was feeling the need to get a wriggle on (not that there was any pressure from them; it was my perception).

We were getting the endorphins going left, right and centre. I was still riding my bike and going for walks; my parents even went out, saying to us, "Go on, have vigorous sex!" (This is really not what you want to hear from your church-going mum.) I even went for a jog!

The sweep didn't work, as the cervix hadn't moved forwards, so the midwife took my blood pressure and because it was high, she said we should call the hospital. After being monitored twice previously during the final month, I was told it wasn't high enough for doctors to be worried, so off we went back home. After supper that evening, I felt strange. I wondered if I had possibly over-eaten, but I knew something wasn't right. I took my blood pressure again and it was much higher than before, so back to the midwives' office we went, and this time we were sent straight to hospital.

My intention to labour at home and have a water birth were off the table: after 41 weeks you are no longer under midwife care. I was monitored overnight. My blood pressure remained high but stable. My mum has had a brain haemorrhage so we didn't want to ignore the high BP and I was happy that baby and I were being monitored.

The next morning, which was a Thursday, I was given a Foley Catheter in the hope that it would encourage my cervix to move and create some dilation. Was it invasive? Possibly, but we weren't at the induction drip stage yet. I should say that at this point, I knew after having read the statistics that my baby was not coming out naturally without drugs. However, I was confident that this birth was going to be great whatever. After living with polycystic ovary syndrome, we never even thought we would be in this position, so the end result was going to be worth it regardless of the process we would go through. Thanks to Suzy for the breathing and affirmations.

Roll on Thursday night: Matt had gone home for some sleep (lucky him) when my surges started. The catheter was doing something, at least. By 2am on Friday, a giant dutch midwife with the longest fingers ever came to see me and check on my progress. My cervix was still being backward in coming forward but she could at least touch the membranes. Sweep done, catheter removed: time for me to get rest. Another midwife comes to visit in the morning, and they decided they wanted to induce me. My BP wasn't improving and they wanted baby out.

We got moved to a birthing suite and I was hooked up to monitors. I could still move around a bit, sitting, standing, squatting. Then it began. I wasn't prepared for the strength of the hormone induced surges, nor was I aware that there would be no let up or break between them. My first thought, "What the heck?!" was followed by Suzy's voice in my head: "Breathe, breathe, breathe." I had affirmations running through my mind the whole time, and at one point Metallica's Enter the Sandman started blaring out, with Matt beaming at me. We were having such a good time. No, it wasn't what we had as our first choice, but we were going to meet our baby boy or girl by the end of the day!

At 6cm I requested an epidural, thinking I would get a reprieve. I'm a potty mouthed swearer, but through the entire process the one and only time I decided to swear was when the anaesthetist gave me the pre-epi prick. I shouldn't really have called him a MoFo: firstly, he and his assistant were both hot, and secondly I would be seeing him later. Oops!

The surges came thick and fast after that. I thought the epidural would help or stop the discomfort, but it just changed it. I went on the bed, Matt holding one of my hands palm open, my other hand stroking the bed bar up and down in time with my counting. Anyone in room 11 will have the shiniest bar on the bed! It was working, we were doing this and loving it.

The midwife came back into the room to examine me: 8cm, woop woop! However, baby's heart rate was not recovering well. They did a blood gas test on baby's head, and the results were right on the cusp of their preferred readings. The midwife and nurse started discussing something in Dutch. I said, "You're preparing for a C-section, aren't you?" They told me that they always like to be prepared. The second test was done and the tempo shifted, gears; it was clear that they wanted to get baby out. They knew what my birth wish had been and we were so far away from it that they didn't want to tell me. I was absolutely fine with the change of plan; I even said, "Brilliant, let's go." For us, it was all about the journey to meet our Boo Boos, who at 8:12pm on Friday 15th May was hoicked out of the birth canal, covered in poop and held above a blue screen to be shown to us: a beautiful baby boy weighing 3550g, not the behemoth we had been told to expect.

William had arrived, our miracle, the baby we thought we would never have, on a day that was so happy, enjoyable and mind-blowing due to the work we put in with The Calm Birth School. I want to say to any woman preparing to give birth that even if your birth wishes don't happen, you can be in control of you and your reactions to the unfolding situation. TCBS, we couldn't be more grateful for all your support.

Wave Breathing

You will use wave breathing when you are experiencing a wave or surge (otherwise known as a contraction).

How to do it
The central idea of wave breathing is to keep both the inhalation and the exhalation even. Breathe in through the nose to the count of seven and out through the mouth for seven.

The role of this breath is to work with the upward motion of the uterus as it rises, and then to send your breath down to your baby and your womb, while relaxing. Simple!

However, don't be fooled. For you to move instinctively into that space of deep breathing and relaxation when you experience a wave, you need to have practiced it often so that you are used to it.

Please do not worry if you are unable to keep the breath even for a count of seven to start with. Work with what ever feels most comfortable for you. Perhaps you'll start off counting to four and once that feels good extend it to five. The main point is you start to feel comfortable slowing your breathing down and taking control of the flow. This will help you immeasurably during labour and birth.

> **When to use it**
>
> My suggestion is to practice this type of breathing for five full minutes every morning. If that means setting your alarm five minutes earlier – do it. It's such a great way to start your day and will leave you feeling great as well as preparing you for your labour day, when you'll be using it during each surge you experience.

Chapter 5: Taking Control

Do the Work

As you can probably tell by now, I like to say out as it is – no messing around. I've explained to you that hypnobirthing isn't a magic wand, but you also know that it is one of the best things you can do to give yourself a great foundation for making your birth as easy, comfortable and as positive as possible. (Yes, I will keep repeating this over and over. We have to make sure the old fear-based message is well and truly erased from your mind!)

By the time you have finished reading this book you will have all the tools that I have shared with thousands of women around the world who have created amazing births, the kind they're happy to share with anyone who'll listen! However, reading the book and listening to the MP3s isn't all there is for you to do. You have got some work to do in order for you to experience the type of positive birth you've been reading about. Actually, make that a lot of work. Yes, you get points for showing up and reading, but that is just the beginning. The time has come for us to make an agreement. This may or may not go against all of your sensibilities, but it doesn't matter. Wherever you are right now, I ask that you suspend any and all beliefs that have caused any doubt about your ability to create a really positive birth.

As a demonstration of this, your task is to write out the following and then say it aloud:

I know I can create an amazing, positive birth.
I will immerse myself in positive birth stories.
I will imagine my positive birth with feeling.
I will do my daily breathing exercises.

This is important to me because _____. (Complete this sentence)

And once you have completed that exercise I would like you to do the same with the following sentences.

I am creating an amazing and positive birth.
I am immersing myself in positive birth stories.
I am imagining my positive birth with feeling.
I am doing my daily breathing exercises.

This is important to me because _____. (Complete this sentence)

Changing the sentences from 'I will' to 'I am' takes things out of the future into the present tense – giving you no excuses to hide and letting your brain know that this is happening.

Don't do this exercise later: do it now, even if it's on the back of an envelope. Don't think it in your head. Actually write it down, and say it out loud.

It doesn't matter if you feel a bit silly saying these words. It really doesn't matter. What's important is that you are fully invested and believe in your body and your baby's ability to do the job they were designed to do, and if speaking some affirmations out loud helps you with this, then it's twenty seconds well spent. If you're already feeling confident, relaxed and committed, this is a worthwhile exercise in reminding yourself why it's important to you to create a positive experience. It's a win-win either way. If however you're feeling a bit anxious and awkward, now is the time to leave Ms Cynical at the door.

Remember the way the brain works. When you *believe* something to be true — even if it's not actually true (in this case, that birth is painful and difficult and it's only fun once I get to the end bit), the subconscious will do its very best to prove you right and prevent you from experiencing this unsafe event. In practice this takes the shape of slowing down or even stopping your labour as it tries to stop you from giving birth, in order to keep you safe. This is obviously not what you want to happen! If you're in labour, you want baby to come out, not to stay in, and delaying this process usually results in more discomfort and medical intervention that might not otherwise be necessary.

The only way we can change the subconscious patterns of belief are to know and *believe* there is an alternative to our engrained thought patterns.

What is a belief? Essentially, a belief is a thought that we have repeated to ourselves many, many times before that we have categorised as "real." By repeating your thoughts about your ability to create a calm and positive birth experience tens, hundreds or even thousands of times, you start to form a new belief for your subconscious to follow. This allows you to look forward to your birth and as you mentally rehearse your birthing experience and your ability to navigate all or any of the situations you face with ease, this becomes your reality on the big day.

We embed new ways of thinking and responding to situations through constant repetition of new actions or thoughts. Think back to when you learned to drive; when you first started learning, you didn't think there would ever be a time where you could have people talking to you in the back seat, whilst having the music on, while navigating a roundabout. Every ounce of concentration had to go into using your mirror-signal-maneouvre process, and you had to practice again and again (and again). If you're not a driver, perhaps you remember learning to ride a bike; it took a lot of practice before it became instinctive for you to mount and ride a bike with ease.

You can turbo-charge the process of embedding these new thought patterns in your mind by adding emotion. Generating positive emotional responses within yourself alongside the thoughts about birth and specifically how you are going to respond on the day will speed up the process of these new thoughts becoming beliefs.

So, how exactly are we going to do this? First things first: if you haven't yet said the above affirmations out loud yet, stop reading right now and recite them. This simple act is symbolic: you are sticking your marker in the sand and declaring, "I'm taking control now." As you move through the course you will also learn how to let go of everything you can't control.

Here are the statements again. If you read them before, here's another opportunity to do it again:

I am creating an amazing and positive birth.
I am immersing myself in positive birth stories.
I am imagining my positive birth with feeling.
I am doing my daily breathing exercises.

This is important to me because _____. (Complete this sentence)

Once you've declared this — and meant it — the other thing you can do is to start listening to The <u>Calm Birth School</u> course daily affirmations. There are written versions available for free at <u>www.thecalmbirthschool. com/bookbonuses</u>. Start reading and reciting them every day. If you're feeling skeptical of the 'war cry' above, I strongly suggest that you make the time to write out and say the above 'I am' affirmations every day at least ten times, before you listen to the recorded affirmations. Continue to do this daily until you really believe in the statements and you are able to dive straight into the recorded affirmations without any preparation.

For some people the words will resonate and hit home right away; for others, it won't feel real until the day they go into labour. Neither is better or worse. All you have to do is commit to getting there.

Once you've have listened to your affirmations, decide which three or four
you love the most. Write them down and post them around the house, on places like the bathroom mirror or the back of the front door so you are continuously reminded of the calm and positive birth you are creating.

Language

The language you use and are exposed to is so important. The words we use, and the contexts within which words are said, have the power to evoke huge emotional responses. A kind word can make the stress of a horrible day fade away in seconds, whilst a person who knows you well might be able to get your blood boiling by stating something that to anyone else would not even register. For example, someone might innocently say to a pregnant mama-to-be, "So are you going back to work once you've had the baby, or are you just going to be a stay-at-home mum?" The word 'just' is a completely inoffensive word, but in that context, it's enough to trigger stress in many women as being a full-time mum could be perceived as being less worthy or important than going back to work.

Words create imagery in your mind which in turn trigger emotions; every thought we have creates a chemical and physical response in the body. That's why it's important to minimise the use of language that has negative connotations to help lessen the stress response, no matter how small the stress level might be.

The Calm Birth School Glossary

Below you'll see two columns: on the left, a list of birth-related words that are commonly used in our society; and on the right, the words I suggest you using from this point onwards. Notice how it feels to say or hear the word in the left column compared to the one on the right.

You'll also notice that throughout this book I have used the terms on the right, for all the reasons discussed above.

Instead of saying	**Try saying this instead**
Contraction	Wave or surge
Pain	Discomfort, sensations or pressure
Birth canal	Birth path
Pushing	Bearing down
Complications	Special circumstances

Birthing Horror Stories

This one is simple: <u>do not listen to them!</u> The "I am doing you a favour by letting you know how terrible birth can be" so-called 'advice' doesn't fly with me. For all of the reasons I discussed above, someone else's experience comes from their perspective and can never be repeated by anyone else, so to plant the negative seed of expectation because of their own personal journey does not serve you or your baby. So from hereon in, you have express permission (actually, this is an order!) to get friendly with your inner child, put your hands over your ears and sing the 'La, la, la, I'm not listening' song.

If that isn't quite appropriate, stop the conversation politely, and firmly tell the storyteller you would prefer to catch up with her birthing story after you've had your own experience. You'll feel great once you have set a clear boundary and you won't be adding any more unnecessary negative stories to your database. If you find setting boundaries challenging, this is a great opportunity for you to start to tap into that protective maternal energy and use it to take care of yourself and your baby.

Release and Let Go

Pregnancy and the thought of birth can take some of us by surprise. We can be so confident in other areas of life but when it comes to our pregnancy bump we get 'worryitis'. If you are already prone to worrying, the unknowns of pregnancy can simply add to your current anxiety. This is not bad or wrong and I've seen it many, many times before, so if you fall into this category, please do not beat yourself up for it.

The Release and Let Go technique will help you put those worrisome thoughts to bed so that you feel as chilled, relaxed and happy as possible, instead of being tense, uptight and nervous. And the great news is that because worrying is simply habit you have learned over the years, anything that can be learned can also be unlearned. Sometimes it takes a bit of practice, particularly in the beginning when it often feels easier to do what you have always done which in this case is let yourself worry, but you have choices available and it's time to take charge.

If you notice yourself getting into a worry loop, your first step is to simply recognise it. When you do recognise it, *do not beat yourself up*. Acknowledge what you're doing and try to have a laugh at yourself and lighten everything up. You can even say to yourself, "Look at me, I'm doing it again!" This is important because by noticing and commenting light-heartedly on what's happening, you immediately interrupt the autopilot loop.

Once you have become aware of the worry, ask yourself, "Do I want to feel worried/anxious/angry [or whatever the feeling that goes with the worry] or do I want to feel better?" Sometimes you won't actually want to feel better right away. I'm sure you can think about a time when a partner or a friend has annoyed you and you felt completely justified in staying in a mood with them, because you were not quite ready to move on from your worry or frustration. If this is you that's fine, but remember to ask yourself the above question, as it will put you back in control of your process rather than at the mercy of it. Acknowledging that you are choosing to stay in the less positive space is the first step in truly moving beyond it.

If and when you do make the decision to feel better, release the thought and think about something that makes you smile, such as something or someone you appreciate or that makes you feel a little better. You could imagine lying in a lovely warm bed, or if you're being kept awake during the night, perhaps you could think about how great it feels to be on a beach with the sun on your face. For some of us, thinking about a lovely piece of chocolate, or a nice memory, or someone we love will do the trick. Once you have found your 'sweet spot,' keep reaching for a better feeling or thought until you are able to sustain that good feeling. You don't have to go from feeling terrible to feeling amazing for this to work. Any slight improvement on your mood or distraction from your worry even for a short period of time is a win. And as with everything, the more you do it, the easier and more instinctive it becomes.

The first few times you try this technique your mind will keep bouncing back to the negative place and the whole concept may feel disjointed. However, success comes with continuing to choose to focus your attention on a more positive thought – which works most effectively when the new thoughts are completely unrelated to babies, pregnancy or giving birth. Although this isn't easy, it's worth it.

This exercise is like any other workout: it takes time to get your positivity muscle working harder than the automatic Negative Nelly who is so used to her job that she can do it without thinking. Once you've got it though, what you'll notice is that by regularly practicing and reaching for a more positive feeling, you will instinctively become more positive about your birth and life in general.

To sum up, your new three-step process to move from worry to chill is, in the following order: *acknowledge, release* and *reach for a better feeling thought*.

Chapter 6:
Where to Give Birth

Creating a Positive Birth Environment

Now that we've covered a lot of the psychological groundwork so that you can create an empowering attitude towards pregnancy and birth, we can turn to the more practical details.

Choosing where you're going to give birth will hopefully be a fun and interesting process. I encourage you to weigh up the pros and cons of all of your options. This chapter will guide you through the various choices available to you are, and will empower you to find the best solution for you.

You might instinctively know that bringing your baby into the world at home feels the most natural option — and then again, you might read that sentence and think, *'You have got to be joking!'* The beauty of The Calm Birth School approach is that there is no right or wrong place to meet your baby for the first time. The most important thing is, in the words of the amazing childbirth educator Sarah Buckley, that you feel "private, safe and unobserved." When you look at the theory behind that, it all makes perfect sense.

Whichever way you choose to slice and dice things, we are mammals and there are certain traits common to any mammalian birth. If an animal is diurnal (awake in the daytime) they tend to go into labour at night, when things are quiet and the daily hustle and bustle is done. Have you ever seen a cat give birth? If so, you've been very lucky, because cats will typically search out the quietest place in the house so that they are unlikely to be disturbed. If you do manage to spot mama cat and try and take a look at what's going on, she will tend to move away to find somewhere she can birth peacefully, without her environment being disturbed. You might also notice that a number of the birth stories throughout this book are from women whose labour either started or took place during the night.

Our genetic make-up means we are programmed to want exactly the same thing. However, few of the more commonly used birthing environments in the modern western world offer us a space where we feel a sense of privacy or even quietness. And unless you choose to birth at home, this certainly won't be a given. So what can you do?

To create a positive birth environment wherever you're choosing to birth, I recommend the following four step process:

Step 1: Know Your Options

Home birth
If you are enjoying a low-risk pregnancy and all the signs indicate that your baby is healthy and happy, you may choose

to opt for a home birth. NICE (the National Institute for Clinical Excellence in the UK) has stated that for second-time mums, home births or low-tech midwife-led units provide better birth outcomes in terms of safety and lower rates of intervention than hospital births.

In attendance you will have either: a community midwife, who you may or may not have met before; a caseload midwife, who you will have grown very familiar with during visits throughout your pregnancy; or a private midwife. Being in your own environment, with a familiar care provider, helps to promote the optimal environment for giving birth.

On-site midwifery-led unit
For many low-risk women, the middle ground is to give birth at a midwife-led unit on a hospital site. Choosing a place where midwives are familiar with natural births and understand that you are looking for labour to progress at its own pace, with the knowledge the hospital is close by. This can be the ultimate reassurance for a mother-to-be, helping her to feel at ease should special circumstances arise.

Birth centres
Birth centres are very similar to midwife-led units. They are run by midwives and aim to create the same type of feeling or environment as being in your own home, or even better a spa, in the same way as onsite centres provide. The main difference between a birth centre versus a midwife-led unit is that they often stand alone, i.e. not connected to a hospital. Once again, these are appropriate for women who are experiencing low-risk pregnancies. If you're based in the UK, there are both

private and NHS-run birthing centres available at the time of writing. If you are based in the US, many birth centres are currently covered by medical insurance.

Labour ward

The labour ward is the number one option for women who are experiencing more complicated pregnancies. If your labour and birth progress without issue, your care will be predominately midwife led. But the labour ward offers consultant/obstetric led care. Should you request stronger pain relief like pethidine or epidurals then generally speaking, these drugs can and will be administered here.

Katie's Story

My waters broke at 6.30pm on 40 +3 in my kitchen. I thought it had happened earlier but there was no denying this! I ended up eating my dinner sitting on the loo, then called the midwife on call who said she'd come when I couldn't talk through surges. We went for a little walk and the surges were very manageable. We put on TCBS affirmations and played them on repeat for most of the night. By 11pm, we called the midwife and my hubby blew up the birthing pool. The TENS machine was a waste of time; it was really only the breathing and concentration work that helped. I imagined waves every time I breathed through a surge. The in-between time was such a wonderful relief. From the time of getting into the birthing pool onwards, time went so quickly: I couldn't believe it when the sun came up!

The surges became really intense once the early morning came. The only description I can give is that they were like the power of thunder; I experienced a really strong juddering that took my breath away. At 2am, things started to slow and the midwives suggested I get out of the pool for a bit and have a rest. I dozed on the sofa through surges until eventually they had to give me a catheter to remove the blockage to baby. By that point, I had been going for about 12 hours and the midwives were keen to get things going, so I ended up walking up and down the stairs and even doing lunges which aren't fun at the best of times, especially having done zero exercise for four months! I didn't do the wave visualisations anymore; by this point, just breathing through the surges was all I needed.

Eventually, they asked me to actively push because of the time and my increasing heart rate. My husband held me as he sat on our sofa, and I gave birth to our gorgeous little girl Eliana at 9.37am on 18th August. All the way through, she had a perfect, consistent heartbeat. She was so calm. The midwives commented on how well me and hubby worked together as a team and how I managed to literally breathe her out and cause minimal damage. I am so, so grateful to TCBS and everything we learned with you, and how accessible it was to my husband too, who was an amazing birth partner. I never thought a birth like this could be possible but it really is true! Thank you so much.

Step 2: Start Asking Questions and Investigate

Once you have a sense of where might be the right place to give birth, you can start investigating in more detail. It's vital to give yourself permission to ask questions. If for example a labour ward feels like it might be the safest place for you to birth but you would still like to birth naturally, I would advise you to:

- Look up what their statistics are for natural births vs unplanned/emergency C-sections.
- Find out whether they support natural labours and births.
- Investigate what their general policy is for a woman who is on the borderline for gestational diabetes.

These are all important questions that, depending on the responses, will either make you feel reassured this is the right place to give birth, or not. There is a fantastic article by one of our Australian colleagues Tina Ziegenfusz aka The Glow Doula, which is definitely worth a read: 6 great questions to ask when choosing your care provider.

Step 3: Get to Know Your Environment

If you're going to be birthing in an unfamiliar environment, ask to be shown around. This helps to reassure your subconscious on the day that there is nothing wrong because this is a place you have already been to before. If you think about it, before getting pregnant, when would you be most likely to visit the hospital? When you were feeling sick or injured or when someone else was sick or injured. This isn't exactly reassuring for your subconscious mind! Therefore, if you're choosing to birth in a hospital, getting familiar with the environment and the care providers during a time when you're feeling relaxed and confident signals to your subconscious that there's no need for you to either consciously or subconsciously be on high alert once the time comes for you to visit when you are in labour.

While it's completely normal for your labour to slow down when you move out of the familiar home environment to go to the hospital or any new environment (in order for the subconscious to double check there is no imminent danger in store), you can help lessen the impact if you get familiar with where the action is going to happen beforehand.

Step 4: Make it Your Own

I mentioned earlier that the most important things for you to feel are private, safe and unobserved. Mark Harris, author of *Men, Love & Birth*, talks about creating the same type of environment for giving birth in as you would for making love in. Bingo!

For the vast majority of us, the place where we want to get most intimate with our nearest and dearest is a place where the lighting is conducive to making us feel like a goddess. LED strip lights and fluorescent tubes don't really cut it! We'd lose the person that wants to keep 'popping in' every half an hour or so to check your progress and marking you out of 10 with a clipboard in hand?! That is neither the optimal environment for making love or birthing babies.

Regardless of where you birth, you want to create an environment where you and your birth partner feel as private as possible and completely safe. I often have clients include a sign for the outside of their hospital room door stating "Hypnobirthing couple. Quiet please and please knock before entering." While you might be cringing at the very thought of doing this, I urge you to put your concerns to one side and remember that you only get to do this once. So if a sign on the outside of the door is going to remind your care providers to speak in lowered tones and of the type of language you'd like them to use, then it is worth it. These seemingly small shifts are really important in helping you to create a birth experience where you feel calm, private and respected.

Think about your birthing environment as a kind of love or birthing nest. If you're birthing at home, what would be good to have on hand to make your experience as comfortable as possible? If you're birthing outside of the home, what can you take with you to help you feel relaxed and happy, bringing with it the familiarity of home? This is important because home will be the place where the vast majority of people will feel at their most safe, so recreating that sense of security where ever you're birthing can only serve you.

Here are some ideas previous clients have share to make their birth place feel as comfortable and as homely as possible - make sure you include them in your list and let your birth partner know exactly what you want and need on the day (more on this in chapter 10):

Fairy lights – always top of the list!
LED candles
Fluffy towels for post-birth
Pillows

Room spray/face mist
Pictures of loved ones
iPod docking station for music or MP3s
Handheld fan
Massage oils
Portable blackout blinds (also great for once your baby arrives)

Certain objects will trigger happy and comforting memories for you, so start noticing which items you can consciously and deliberately create positive associations with in your home environment that you can then take with you into the birthing room. Some of my clients create their own birthing playlist, which is a great idea. If you want to do this too, I recommend taking the time to listen to your playlist and practice your breathing techniques while listening before you go into labour. Listening to this out loud means that your baby will enjoy it too!

Others like to spray their rooms with their favourite scent. Lavender is very calming and soothing, and spraying it around your bedroom in the evening before listening to your MP3s is a great idea. By the time you go into labour. you have a really strong association with your chosen scent, which you always smell when you are feeling relaxed and at ease.

Chapter 7: Choosing Your Care Providers

Things to Consider

Choosing who is going to support you as you welcome your baby into the world is an important part of your preparations. You want care providers who are going to make you feel special, cared for, understood and respected. Having these core needs met will contribute to how you feel on the day you give birth. The more at ease you feel with the people who are going to be supporting you, the more likely it is you will be able to create a positive birthing experience.

At the time of writing, if you choose to work with the NHS in the UK and are not with a caseload team (more on what that is shortly) you will meet with a member of a community midwifery team. During all stages of care, including the day you give birth, it's important you like and feel safe with the team you are working with. If you do not feel positive about your midwife or consultant, you are entitled to ask for a different member of staff to work with at any stage.

Some people think it will be too embarrassing to ask for someone else, or don't want to be known as 'that difficult hypnobirthing couple'. They might worry that there won't be enough staff to accommodate their wishes. A phrase that has always served me well when working with my face-to-face clients is, 'What other people think of your is none of your business.' Certainly, when it comes to staffing, *it's not your responsibility*. The only thing you need to concern yourself with is feeling safe and comfortable, so you can feel positive during your pregnancy and crucially during your birth, where one of your primary tasks is to help your body to work efficiently. If a member of the care team causes you to feel tense, anxious or even angry, this can slow your labour down as your subconscious attempts to protect you and your baby from the perceived threat in your environment by triggering flight, flight or freeze mode, the consequences of which we covered in some detail earlier in the book.

The main point here is that *you only get to do this once*. So, what do you think its more important: putting up with support that is actually hindering the natural process of birth, because you don't feel at ease? Or ensuring that every single person in your space is able to serve you in the way that is going to be most conducive for you to enjoy the birth experience you desire?

Managing your birth environment is one of the key roles of the birth partner, which we'll cover in detail in chapter 11. In brief, when your birth partner is in tune with you, they can sense anything causing you to feel stressed. The best way for your birth partner to heighten their awareness of how you are feeling is by being present and paying close attention to you.

Part of the birth partner's job is to respond to the subtle and sometimes not so subtle changes in your body, noticing whether tension is creeping into your posture or whether you have a clenched jaw or worry lines across your forehead. If these things are being triggered by a particular person, that person should be asked to adjust their behaviour, or asked to leave your environment without question.

Some of my clients have a trigger word or sign to let the birth partner know they need to do something, whether it's a look, a wink or a completely out of context whisper of the word onion. They know they are needed.

Who Is Going to Be in the Room With You?

There's an old wives' tale you can put an extra hour onto your birth for every person that's in the room. Now whilst I don't think this is true, feeling private is important. However, rather than focusing on the number of people in the room, we're going to focus on how safe you feel.

What can you do to help yourself feel at ease? Simple: Have your dream team surrounding you.

Primary birth partner
Ideally, this is a person who knows you really well. They will have been supporting you through your pregnancy and you will have worked through this book together, understanding the techniques and discussing what will be most helpful in helping for you on the day.

Your birth partner is your advocate. They are by your side and *on* your side, and should be the first port of call when managing your environment, so you don't need to worry.

It's important you feel confident your partner can do their job effectively, as in an ideal world I would like you to be as disengaged from rational and logical thinking during your birth unless absolutely necessary. This is because the minute you feel the need to get involved with stage managing your environment, the more likely it is for you to be unable to stay in the state of mind that provides you with both the emotional and physical capacity for calmness that you need to birth more quickly and comfortably. See chapter 11 for a more in depth look at the role of the birth partner.

Doula
I love doulas! Doulas help to build women's confidence, and as an expert in childbirth, will help you to feel relaxed, more comfortable and safe whilst barely having to say a word. They offer unconditional and consistent support to you and your birth partner, providing better physical and emotional birth outcomes for mum and baby. Doulas can also be of great help post-birth too. Be aware that most doulas specialise in either being a birth doula or a post-natal doula.

In 2012 Hodnett *et al* published an updated Cochrane report on the impact of continuous care – this is care from the same care professional throughout your pregnancy and birth. The findings are pretty unequivocal:

- 31% decrease in the use of Pitocin (synthetic oxytocin used during induction)
- 28% decrease in the risk of C-section
- 12% increase in the likelihood of a spontaneous vaginal birth
- 9% decrease in the use of any medications for pain relief
- 14% decrease in the risk of newborns being admitted to a special care nursery
- 34% decrease in the risk of being dissatisfied with the birth experience.

Why is having continuous care so powerful? Because it makes you feel safe and secure. Unfortunately, midwives are often unable to give you the type of continuous support that would be ideal as they are working with multiple clients and have huge amounts of paperwork to fill out whilst you are birthing.

You might be wondering at this stage, "But what about my partner?" A doula is not there to replace your partner at any step of your journey, but often what a doula is able to bring to the table that your birth partner may not be able to is the experience of being an expert in birth. Doulas have been trained to support you both, noticing the slightest stresses and providing you with the reassurance that you are going to be just fine in a way a birth partner may not be able to do quite as effectively. Having your birth partner in the room with you, if you are both keen for that to happen, can be magical and nurturing for both parties. However, the statistics are clear: women who are supported by doulas (both with and without birth partners being present) experience better birth outcomes.

If you can't afford or don't want a doula, that's fine too. Millions of women give birth every year without doulas and have fantastic birth experiences. I have given birth twice without a doula very positively. So as always, tune into what feels good and if a doula isn't what you feel you require, then it's all good. If you do want one, but are not sure finances will stretch, there are lots of doulas who do pro-bono work and even students who are learning their craft. Stay open to finding a great alternative that will support you for a significantly lower investment because they are available.

Named, Private or Independent Midwife
Some trusts in the UK offer named midwives as part of the NHS, which is fantastic. Having the opportunity to build a relationship with the person who will be assisting you during your birth helps you to feel more calm and at ease with giving birth. However, for those of you who don't have the option of a named midwife, Private or Independent Midwives can be a great alternative, since they will provide you with all of your antenatal and postnatal support — and if you're not birthing in a hospital, even the delivery of your baby.

Best friend/Sister/Mum/Mother-in-law
Whilst some of you will cringe at the thought of your mother-in-law being at your birth, for others it won't be out of the realms of possibility.

Neither is right or wrong. The only thing you need to consider is whether the people you're thinking about are up for the job? Are they able to put their own needs aside and put you and your beautiful new baby at the centre of their focus while giving you the kind of support you most want and need? If you feel confident in having a family member or friend as your birth partner, then great. If you're not so sure, perhaps put them on the 'first to know when baby is here' list.

Holding the Space

Whoever you choose to support you during your birth, the main job facing them is to hold the space for you. They will best be able to do that if they feel prepared and calm and at least have an understanding of what you have been learning in your preparation. Inviting your birth partner to practice the breathing techniques is not only for your benefit; it will be great for them too, helping them to stay calm and relaxed during your birth and ensuring they are not sending out any subtle subconscious vibes to you that there's something to worry about.

Chapter 8: Communicating With Your Care Providers

Get Specific

One of the best things you can do to create a positive pregnancy and birth experience for you and baby is to make sure all of the services and advice you receive are specific to you and your situation. This means asking questions of care providers in a way that provides you with very specific information to your unique pregnancy and birth, rather than accepting general rules of thumb that may not be applicable to you. This is important both in the lead up to birth and on the day.

It can be intimidating to be told by a care provider that you need to take a specific course of action that you either don't understand or feel is unnecessary. Having the confidence to ask why and how certain things are relevant to your situation specifically can make all the difference between creating a positive experience and feeling like birth was something that happened 'to you'. I will share a specific set of questions you can use later in this chapter.

When you practice asking questions during pregnancy such as, "What makes you think that about me in particular?" or "What would best suit me in this situation?" you will gain clarity, allowing you to make informed decisions about how you would like to progress. Having a sense of control is a key barometer for wellbeing. This need is particularly heightened during pregnancy and birth.

While sometimes you may conclude the best thing to do is relinquish control and move forward with your caregiver's advice or opinion, when that decision is made by *you* (armed with facts that are relevant and specific to your pregnancy), you remove doubt and confusion and know that the decisions *you made* enabled you to get the right birth for you and your baby on the day.

Use Your BRAIN

In order to access an experience of feeling calm, empowered and in charge of your birth, I invite you to lead and direct the conversation using the acronym BRAIN to get clear on what is being requested or suggested by your care provider. You can then make your decisions in an informed way, and as a team.

B: BENEFITS
What are the benefits to me and my baby of moving forward with the suggested action?

R: RISKS

What are the risks to me and my baby of moving forward? What are the risks of not moving forward?

A: ALTERNATIVES
What are my other options?

I: INSTINCTS
What is my instinct on this, my gut feeling?

N: NOTHING
What is the immediate risk if we choose to do nothing for the next 30 minutes/one hour? (Be specific about the time period.)

It is also a good idea to include in your birth preferences that unless there is an immediate medical need for either you or your baby, you and your birth partner would like to be given the time and space needed to discuss your options.

When it comes to your pregnancy and birth, clarity is king! So, remember to use your BRAIN and give yourself the opportunity to make informed choices. Ana's story below is an example of how even when birth unfolds in an unpredictable way, you can choose to feel calm, empowered and in charge.

Birth Preferences

You will communicate with your care providers via your birth preferences, commonly referred to as your birth plan. I deliberately don't use that term when talking about your birth choices, as plans can often feel more definitive - and from the stories you've already read so far in the book, you'll know that nothing is definitive when it comes to birth! People making plans for weddings totally makes sense. It would be completely inappropriate for most people to have wedding preferences: "If you don't have enough salmon for the wedding breakfast, don't worry, we'll have beef." However, birth cannot be neatly packaged up in the same way as a wedding can.

Be wary of veering too far in the opposite direction, however. While there are no guarantees in birth, the idea of not completing a birth plan at all because nothing ever goes to plan (a comment that I hear frequently) completely misses the point. After talking to the doulas over at Doula UK, I would recommend that when you discuss your birth preferences with your birth partner, you take an A, B, C approach. Think about what you would like to happen during a straightforward birth (Plan A), how you would like things to proceed if you need some assistance (Plan B), and how you would like to be supported should you end up in a situation where an unplanned caesarean is the best course of action (Plan C).

Taking this approach means that while it is important to spend your time and energy focusing on your ideal birth scenario, you have also taken into consideration the practical elements that may arise should you not have a straightforward delivery. This serves you, your birth partner and your baby far more powerfully from an emotional perspective when you are able to navigate the 'unexpected' from a place of preparation, rather than having to make knee jerk decisions on the spot. And while you can never plan for every eventuality, understanding the importance of using your BRAIN will also help you immensely to turn difficult situations into infinitely more positive ones.

For ideas on what you to include in your birth preferences, please use the following link to download our birth preferences planner here.
www.thecalmbirthschool.com/bookbonuses

Ana's Story

I am so excited to share my positive birth story with you. We are out of hospital and have done some settling in at home, and I can now switch the computer on whilst my two little cherubs sleep - go, go, go!

It all started when I was taken into a delivery room with two lovely midwives who had read my birth preferences and said they would do all they can to help me achieve them. My waters released on Friday morning at 4.25am and the waves began pretty quickly. They grew in intensity over the next 4 hours. I used a TENS machine and my calm breathing and wave breathing, which worked brilliantly to help me stay in control and enjoy each wave as I knew each one was step closer to my babies. Unfortunately I hadn't dilated so I was offered the hormone drip - which I politely declined - and we were able to go another 4 hours with the belief that the babies just needed a little more time. Unfortunately I still hadn't dilated so I agreed to the hormone drip.

The surges came thick and fast, and I continued using the TENS machine and breathing techniques along with a mat, bean bag and ball which all helped me relax, flop and go into my birthing zone. I have to mention here that there was no offer of 'pain relief' as I had made this clear in my birth preferences, and I greatly appreciated it because I felt that my care providers had faith in my ability to birth. I used very deep noises as the surges peaked which I didn't expect to do, but it helped massively - as well as counting how many breaths I needed to get me through each wave. This was a saving grace because I knew how long each wave would last for (although some were a lot longer or shorter).

Because my husband had watched the Calm Birth School videos with me, he was an absolutely brilliant birth partner, giving me drinks, space or comfort when needed. When the waves got really intense the best 'pain relief' was actually holding onto him. It must have been the oxytocin! Anyway, after 16 hours I still hadn't dilated and I had a temperature so the doctor (who was brilliant at stopping talking to me when another wave appeared!) suggested increasing the hormone drip.

This was such a hard decision for me because it wasn't something I wanted. Along with saying yes, I knew that I would need an epidural (not once was this suggested to me. I felt in total control). However I knew, using the BRAIN process, that the benefits outweighed the risks: the fact that I'd been birthing for so long with little effect meant that if I did nothing, it could be an extremely long process which would tire us all out.

I asked for an epidural to help numb the intensity of the surges from the increased hormone drip. This is where it gets interesting and I believe it was all meant to be. Firstly, because of the TENS machine and using the TCBS breathing techniques, I didn't feel the epidural go in; secondly, it meant I could have a break from the intensity.

Most significantly, an hour after having the epidural, I was told that twin 2's heart rate was worrying and that it was beneficial to deliver the babies as soon as possible — again the care providers left it to me to deduce that they meant a C-section, allowing me to feel that I was the one in control, not them.

So, after 20 hours of labour where I'd been able to use all of my well-practised techniques, I was taken to theatre where my beautiful babies were born. And do you know what? It was such an amazingly calm, relaxed, light-hearted atmosphere. Everyone was so lovely, informative and respectful. Again, because I knew that my birth could take any twist and turn, I had so many tools in my toolbox from The Calm Birth School, and my husband and I were able to embrace whatever avenue our babies' births took. I can honestly say it was one of the best days of my life. Yes, it was intense, challenging (I had to dig very deep), and long. Yes, unfortunately I wasn't able to have everything on my birth preferences (delayed cord clamping with twins hadn't been done there before and as it was classed as an emergency C-section we agreed that it was best just to get the babies out).

However, I felt empowered, strong, confident, positive and calm throughout. The breathing techniques, positive affirmations and confidence skills I had developed from The Calm Birth School had no doubt allowed me to enjoy the birth of our babies and despite not having them naturally, they were born into a happy, calm, lovely atmosphere which at the end of the day is what we wanted. I felt like I changed that day, not only by becoming a mother, but because it was the first time I had truly committed to something that no one around me had done before, where I really believed in something and I succeeded. I am incredibly proud of myself and can use the skill set

I now have to take me into motherhood. Thank you TCBS for all your positivity and support: we couldn't have done it without you.

Chapter 9: Understanding Estimated Due Dates

Think 'Guess Date' Not 'Due Date'

Let's turn our attention to estimated due dates, or guess dates, as I like to think about them. I totally get it: being pregnant for 40 weeks, particularly if you are one of the early birds who found out as soon as you conceived is a *really* long time! This happened to me with my son Caesar.

But eventually, the day arrives: your magic estimated due date. If you're like the majority of the pregnant population, your ankles have somehow merged with your calves, rings no longer fit on your fingers, you haven't been sleeping well for weeks or months, you're tired of wanting to go to the toilet every hour, only to realise you can barely squeeze out a wee that would fill a pipette and turning over in the middle of the night feels like trying to shift the Taj Mahal. I remember it well. Your EDD shines down on you like a beacon of hope. Not only is it the day you supposedly get to meet your gorgeous little bundle of joy, but the many, *many* other benefits that come with giving birth make you cling onto your due date like a limpet on a rock face.

The problem with this though, is that 96% of us are being led up the river without a paddle. For most of us, the day arrives and… nothing: nada, zip, diddly squat. This is for a number of good reasons. The 40 week due date is based upon Naegele's Rule. This theory was originated by Harmanni Boerhaave, a botanist who, in 1744, came up with a method of calculating the estimated due date based upon evidence in the Bible indicating human gestation lasts approximately 10 lunar months. The formula was publicised around 1812 by German obstetrician Franz Naegele and since then has become the accepted norm for calculating the due date. However, there are more than a couple of massive question marks in Naegele's theory.

Strictly speaking, a lunar (or synodic – from new moon to new moon) month is actually 29.53 days, which makes 10 lunar months roughly 295 days, a full 15 days longer than the 280 days gestation we've been led to believe is average. In fact, if left alone, 50–80% of mothers will gestate beyond 40 weeks. I've always gone against conventional wisdom and insisted that pregnancy is a ten month process.

While your period may have managed to get in sync with all of the girls in the office, the reality is, we all have different length cycles. Even when we know the date of conception there is a five week variance with healthy mothers giving birth to healthy babies. So, if you give birth at 37 weeks, you are not three weeks early and, if you give birth at 42 weeks, you are not two weeks late. You are only postdate once you exceed 42 weeks.

You might be surprised to learn that only 4% of women give birth on their due dates, which is why I prefer to call it a guess date. (Remember the power of language from chapter five?) From the moment you are told your guess date, I strongly advise you to do your best to forget it. If you find that the midwives at your antenatal appointments are slightly obsessed with it, don't worry: it's their job to be. Make your peace with knowing it, but perhaps don't share the specifics with your friends and family. The last thing you want is people texting and asking you, 'Are you still pregnant?', which is beyond annoying. Instead start telling people that you're due around the middle or end of the month (or even two weeks after your guess date if you want to say a date!), to relieve yourself of any additional pressure. This might feel like you're being dishonest, but it's really an act of self-care, so that as you approach and possibly pass your guess date, you're not being hounded with messages and phone calls checking up on you.

Chapter 10: Preparing for Labour and Birth

We've taken a look at the emotional side of birth, but we all know it's a pretty mammoth physical task too, so in addition to practicing your breathing techniques daily, how else can you help your body to prepare? This chapter will walk you through some key areas to focus on as you prepare for your labour.

Pelvic Floor Muscles

The female body is a magnificent thing, perfectly designed to create and hold a baby for nine months. However, I'm not going to try and kid anyone: growing a human being puts a lot of stress and strain on our bodies, even when we come from a natural default setting of being fit, strong and healthy. One of the areas that quite literally takes its fair share of the load is the pelvic floor.

Your pelvic floor is made up of the muscles that run from the base of your pubic bone to the back of your spine. Imagine them shaped a little like a hammock or a sling. They keep everything in place. Most of us won't think about stress incontinence in relation to ourselves – even for a moment – until pregnancy kicks in, when people suddenly can't stop going on about how important the pelvic floor is. Looking after your pelvic floor during pregnancy means that you:

- Are less likely to leak urine after birth
- Are less likely to experience vaginal prolapse (when your pelvic organs begin to bulge into the vagina)
- Will enhance your post-baby sex life.

Techniques

If you would like an in depth look at what you can do to increase and protect your pelvic floor (and avoid those embarrassing little accidents post-birth), check out our Pilates Masterclass with Dr Joanna Helcke, at www./thecalmbirthschool.com/bookbonuses. However, a great starting point for your pelvic floor preparations are the exercises outlined below.

Isolate your pelvic floor muscles by stopping the flow of your wee, just once, when you're on the toilet. Although you may feel a slight tightening of your abdominal muscles, everything else should be free from tension. So no squeezing of your bum cheeks or thighs. Once you have got this, use your muscles to imagine stopping the flow (but not actually while having a wee), holding for five seconds, and then releasing. Repeat the same action five times.

If you really want some serious strength down there, then opt for this second method – "Floor 5 Please, Sir!" Imagine you are going up in a lift, starting at floor one. Squeeze and hold your pelvic floor for around five seconds, then draw up to floor two, holding for five seconds. Draw up once again to floor three until you have reached the fifth floor. Once you're at floor five, don't let go, but squeeze and release repeating the same action all the way back to floor one.

It's never too early to start your pelvic floor exercises, so get going! Once you've had your baby, if you have had a straightforward vaginal delivery, it's advisable to start practicing again within 24 hours. If you have had an assisted delivery, wait until you are passing urine normally before starting up again. For some women this will be straight away, and for others it will take a little longer to be restored.

It's important to do your pelvic floor exercises on a regular basis. I recommend choosing an activity you do at least twice a day and using those times as a trigger to get you 'moving'. So you could use brushing your teeth, or passing a traffic light, in order to make it easier for you to remember to include them within your daily routine.

Perineal Massage

Perineal massage doesn't tend to be a favourite, but I can't understand why! The perineum is the piece of skin that runs from the base of the vagina to your anus. When you are labouring it gets extremely thin, so to prevent any damage and to reduce the likelihood of tearing or episiotomy – especially for first time mums – it is recommended that perineal massage take place from around 36 weeks into your pregnancy, once or twice per week.

Technique

Place one foot up onto the seat of a chair in front of you. Ensuring you're using clean hands, use any vegetable-based oil to coat either the thumb or your index and middle finger. Insert into the vagina and gently massage and stretch the area in a U shape motion. This should be done every day for three to five minutes per day. Alternatively, you could get your partner to do this for you.

Some people feel totally uncomfortable with the idea of perineal massage, in which case I would like to direct you to an excellent product called Epi-no. You can find out more information about it here: www.epi-no.co.uk

Prenatal Bonding

Prenatal bonding is a hugely important part of the process of becoming a parent. From a mother's perspective, it helps you to tune into your baby and your body more acutely, which is invaluable when you are birthing and helps you to feel even more of an instant connection with your baby. For partners, it is an excellent way to create a bond with their babies before birth, which not only helps create a sense of involvement for the partner during pregnancy, but also helps them feel more of an immediate bond with baby once they have arrived. It's not unusual for some partners, particularly dads, to take up to six months to bond with their new babies. The time you spend preparing as a couple for your baby's arrival, can be hugely beneficial to the attachment process and prenatal bonding can help to take things to the next level.

Pre-natal bonding involves acknowledging and interacting with your baby on a playful and emotional level whilst they are in the womb. It's easy to recognise our babies are growing physically within us as our stomachs look fuller every day, but tuning into how our babies are growing on an emotional level is not something we always consider.

While we can start prenatal bonding at any time, the last trimester is when our baby's brain development kicks into overdrive. This is the time when our babies are laying down all of their instinctive patterns of behaviour as they spend 80% of their time in the R.E.M (rapid eye movement) state. R.E.M. is the state of consciousness you enter whenever you are learning something new. Child psychologists talk about what happens between the ages of 0 and 3 being crucial, as our children will spend 60% of their time in R.E.M state, downloading much of what is going on around them, creating the templates they will live their lives by. By the time we are adults we only spend around 20% of time forming these templates, unless we utilise tools like hypnosis, which allow us to enter into this state of conscious, helping us to learn new patterns of behaviour more quickly and easily.

During the last 12 weeks of pregnancy, your baby will learn to look at a human face as opposed to an animal's face when they are born. They will learn how to instinctively grab hold of your finger if you place it in their hand, and they will also know they are able to mimic and copy, which is why if a person pokes their tongue out at a newborn with the first 24 hours, many babies will return the favour. Being able to mimic, engage and hold on are all vital skills for survival as they help to promote bonding and attachment with parents immediately post birth.

Babies communicate with us through their movements, responding to our thoughts, emotions and our external environments, which, as far as I am concerned, are all inextricably linked. Perhaps this is why our mums often report that a baby has been relatively quiet throughout the day will start kicking and stretching like crazy whenever they start listening to their MP3s. I like to think this is baby giving the nod to all of the wonderful endorphins coursing around the body.

This is also why some mums worry about the impact of stress, anxiety and anger on their babies. While it is true being exposed to consistently stressful situations during your pregnancy is not ideal, it's not a good idea to beat yourself up every time you experience a cross word with your partner or find yourself in a stressful situation. In fact, a little bit of a difficult episode is the perfect time to practice your Calm Birth School techniques. Being able to instinctively turn to your breathing exercises helps you return to a feeling of emotional calm when faced with those unavoidable difficult or challenging situations takes practice. So embrace these situations, as it's your ability to tune into that calming space within that is exactly what you are going to need to do when you are birthing.

You are also teaching your baby the same valuable skill. Whilst we all instinctively want to protect our children from any difficult emotions or encounters, real life means they will face these situations regularly. It's likely that at some point during your pregnancy that you will feel stressed or frustrated with your partner. This generates stress hormones, however, when you are able to call upon your Calm Birth School tools, your baby gets to experience the return to calm as your body responds with endorphins when you choose to focus on your breath. The lesson you teach your baby here, as they already start to develop their own emotional intelligence, is that even after a storm things always return to a state of calm. The message is that everything will always be all right, which is a very beautiful gift.

So what can you do to help with this bonding and learning now?

Relax
Every time you access a state of relaxation, listening to your audios and visualising your baby being born, you are strengthening the bond and connection between the two of you as well as creating future memories for your hard drive.

Talk and read to your baby
Baby can start hearing your heartbeat and distinguishing between sounds and voices from around 23 weeks. Getting your partner to regularly talk to or read to baby is a great way to start getting familiar with each other.

Massage your bump
Another great way to get your partner involved is gently massaging your bump whilst talking to you both. Your baby gets to associate your partner with the feel good hormones, endorphins.

Play games
Responding to your baby's kicks can be both reassuring and a great way to have fun with your baby before they arrive.

Sing
They don't care if you sound like Madonna or not, they find dulcet tones reassuring, so sing it out sister!

Music
Playing the same songs so baby becomes familiar is another great helper once baby arrives, in them being settled: they will associate your favourite pregnancy songs with being safe and secure in the womb.

Hazel's Story

I was at week 40 +6 and had a sweep booked at 3:30pm. Feeling positive and calm about the idea that this might slowly kickstart the journey to meeting my baby, I was quite surprised to hear that I was actually 2cm dilated, with a softening cervix and baby's head very low down. The midwife hinted that I wouldn't need inducing the following week; she'd be out soon!

I went home with my mum, who had spent many days with me when Julian was still at work, and had a lie down on the bed whilst on the phone to another mummy friend. As soon as I hung up at 5pm I heard an internal pop/thud and went to the loo. My waters were trickling, and suddenly I had very strong cramps very low down in my body. They were coming every couple of minutes, lasting for around 30 seconds each time.

I did worry that if this was early labour, what would the end be like?! But I lay on the bed and tried to concentrate on breathing while mum called the hospital. I think I even put on my affirmations MP3 - I can't quite remember! The midwife asked to speak to me directly and advised that as my waters were clear I should progress at home. This being my first baby, I really wasn't sure what else to expect of these surges - if these were even surges, as my belly didn't move at all, it was all much lower down - but the midwife could hear how frequently I was zoning out of the conversation and my mum explained that all four of her kids were born very quickly. They advised us to go in for a check up and to make sure my waters had indeed broken.

Mum first called Julian who left work to meet us at the hospital, then rang a taxi. By the time I'd made it down our tiny staircase - having one surge at the top and another at the bottom about 30 seconds later - I felt the cramps move to my bum and the need to push. Change of plan: call an ambulance! The telephone operator asked mum to lie me on a blanket in the living room to check for any signs of baby's head. This is the only time I remember trying to use the 4/7 breath between surges - during them I couldn't focus at all, mainly I think because I was nervous about the idea of my mum delivering my baby (as was she!).

The good news is that a paramedic arrived quite quickly, followed by an ambulance. They said I had time to get to the hospital so I was bundled on and learned the beauty of gas and air. It really helped during the surges as well as over the 15 minute very bumpy journey. Coincidentally Julian met us in the lift on the way up to the ward (not that I was very aware of what was happening). He'd been asked to keep the lift open by the paramedics, and when he looked down at the wheelie bed, he saw me!

I went into a labour room with an out of action birthing pool in the corner, and asked whether a MLS water birth room was available instead. All my ideas of using the water, aromatherapy, a bean bag and ball went out of the window when the midwife examined me. I was by this point fully dilated with baby's head in a perfect position for me to start bearing down during my next surge. Labour definitely didn't slow down when I left home then! I think I was so relieved that my mum wasn't going to be my midwife that my body let go!

The last half an hour I was taken off the gas and air to breathe as much oxygen down to the baby as possible, and was advised to make the most of each surge to push her out. I was on my back (why?!) so swapped to leaning over the back of the bed on my knees, which immediately felt better. Julian was right beside me stroking my back, playing soft music on his phone and spraying lavender on the bed so it didn't smell like a hospital. The lights were dimmed too. My mum was about to leave us to it but she'd been such a big part of the action I asked her to stay. I had a lovely midwife and encouraging trainee so felt very safe.

I can't say I breathed baby out, but the low humming I did whilst holding my breath and envisaging her coming out really helped! I reached down when I felt the head emerge and one last push brought her out and up into my arms. I had requested delayed cord clamping but because I lost quite a bit of blood fairly quickly, after three minutes Julian cut the cord. Similarly I abandoned my ideas of a natural placenta delivery so they injected me to help get it out pronto. I just held my baby in wonder. We had worked together as a team; I'd trusted that she knew what to do, and the 'all is calm, all is well... I am safe' mantra ran in my head too. Each surge brought her closer to me, but blimey, in the grand scale of things, there weren't that many!

I think the **best** *thing about my calm birth experience is how mellow I was pre-birth. Everyone commented on it towards the end. I didn't know what to expect overall but I felt confident in my ability to go with the flow and make informed decisions. Thanks to TCBS, I had the natural, positive experience I had hoped for. And, despite it being super intense, it was a real bonus that it was so quick!*

Chapter 11:
Your Birth Partner

The Birth Partner's Role

When I used to teach hypnobirthing to couples face to face, I would always joke that the birth partner was supposed to be a bit like Batman: the one with all of the tools, techniques and knowledge about how to support you best on the day you meet your baby. Many of the women who take The Calm Birth School video course love the fact that their partners can watch the videos and get a full picture of what they can do through learning exactly what you are learning.

The role of the birth partner begins long before labour. If you haven't been sharing the tools and the philosophy of the book, make sure that you start sharing everything that you are learning. Your birth partner's energy on the day you go into labour will have a profound impact on how you are feeling even if this is at a subconscious level.

During Pregnancy

Breathing techniques
I cannot over-emphasise how useful it is for your birth partner to learn your breathing techniques with you. Your birth

partner likely knows you better than anyone in the room and is the person to whom you have the strongest emotional bond. If he or she is taking long, slow deep breaths and radiating a sense of calm and peace, you are far more likely to pick up on that vibe and adjust your breathing accordingly without them needing to tell you to do anything differently. Breathing with you allows them to guide you to relax and lengthen your breath without any need for a conversation.

They might also, however, unwittingly push your buttons by trying to pass on well-meaning advice. This usually happens because of anxiety, which is the second reason why I advocate them learning and practicing the breathing techniques during your pregnancy: so that they can manage their emotions and thoughts during your labour, controlling any adrenalin that may be coursing through their veins as they witness this miraculous event.

Massage
I have already talked about the power of positive touch. Although you may not want this type of touch on the day you go into labour, if you are well practiced in this during your pregnancy, the positive associations of your partner's touch will help you even if all you do is hold hands. If you get used to visualising birthing your baby whilst your partner is touching you during pregnancy, you strengthen the association between you and your partner when it comes to your baby's birth day.

Practical Preparation

Get your birth partner to practice packing your hospital bag with you during pregnancy and prepare a checklist of all of the items you need, so they don't have to keep asking you where certain items are once you are in labour. Do this even if you're birthing at home, so that your birth partner will be able to gather everything you need.

On the Day

Early labour
The early phase of labour – particularly if your birth partner will be co-parenting with you – can be a fantastic time for bonding, connecting and unifying the two of you before your baby arrives. Use these moments to tune into each other and the fact that life will never be quite the same again. This part of your labour can be deeply, deeply romantic and intimate. When you are doing this with baby number two or more, it can be more challenging to create this space than the first time around. But if your older child or children are present, see if you can make it a family moment instead – it's all doable.

While not quite as romantic, other practical things your partner can do are to remind you to use your birth ball, and making sure your musical playlist is ready and that you have funny films on tap.

Your birth partner is your servant
Your birth partner is responsible for ensuring you are super comfortable in your early-labour and birthing environments, making sure that you have everything you need — although,

if you're labouring during the daytime, it should be business as usual for as long as possible, ensuring there enough snacks, drinks, comfort, your birth playlist and MP3s are on hand.

Oxytocin generator
Your birth partner is there to do anything and everything you need to generate oxytocin. This may be laughter, suggesting going for a beautiful walk, going for lunch and reassuring you or giving your pep talks. Or, it might be kissing, hugging, massage, slow dancing. Anything that is going to raise your mood is great.

At the hospital
The role of the birth partner is to:
- Talk your care providers through your birth preferences
- Manage the birthing environment
- Breathe with you
- Watch out for any build up of tension in your face or body and assist you by remembering to breathe
- Hold your hand and/or providing massage if you desire
- Whisper words of encouragement
- Encourage you to drink fluids regularly and eat to keep your strength up
- Remind you to go to the toilet every forty-five minutes or so: babies do not like passing full bladders
- Be present.

Alex's Story

I'm Alys' partner and very proud father of Clemmie. I know Alys has been in touch with you since giving birth and subsequently her story has gone up on your website. I did a lot of the TCBS course with Alys, watching the videos, practicing the exercises and listening to the MP3s at night.

I thought the content was excellent. Whilst not much of Alys' labour went to plan, I was completely blown away with the manner in which she handled each 'setback'. She was beautifully composed throughout and I really couldn't be prouder of her.

I've avoided using the word 'traumatic' - whilst I would never want to play down the significance of each issue Alys dealt with, I firmly believe that her mental strength and calmness ensured that it was not a traumatic experience. These are characteristics that are in Alys' core but I believe that the TCBS course played a part in identifying and augmenting these traits with the use of the various techniques in the course. I'm grateful for the course being available but also to the additional support you've provided through email and messages. Thanks so much TCBS.

Chapter 12: The Induction Process and Special Circumstances

Remember, the thing that gets your labour going is oxytocin – the hormone of luurve. If your guess date has been and gone and you're getting tense and frustrated because baby doesn't seem to be ready to make an appearance, have a word with yourself and partake in some oxytocin inducing, relaxation-friendly activities.

Continuous Monitoring

You might find yourself in a position where you are invited to have continuous monitoring if:

- You have high blood pressure
- Diabetes
- Your waters have broken and more than 24 hours have passed
- You have had a caesarean in the past
- You are expecting twins

It is incredibly important that you use the BRAIN system and ask for the risks and benefits of being continuously monitored in your specific situation, particularly if the reasons are non-medical such as having had a previous caesarean or you're expecting twins, for example. Statistically, women who have continuous monitoring, which forces them to lie on their back, are more likely to end up requiring some other form of intervention.

Always find out what your options are when it comes to monitoring and remember *you always have choices*.

Pre-Labour Interventions

The reason so many people spend time discussing the arbitrary forty week estimated due date is because this date plays a big role in determining whether a woman will be advised or offered a pre-labour intervention, such as one of the following.

Stretch and Sweep
A stretch and sweep is where your care provider will use their fingers to massage the neck of the cervix, in the hope of stimulating the uterus and kickstarting labour. If this technique is effective labour will start within 48 hours. Depending on the local policy of your care providers, once a women passes her estimated due date she may be offered a stretch and sweep.

The statistics on the effectiveness of membrane sweeps according to the British Journal of Obstetrics and Gynaecology are:

24% increased chance of going into labour within 48 hours;
46% increase in chance of delivering within a week;
74% reduction in going two weeks past your estimated due date.

As with all interventions, a stretch and sweep can only be carried out with informed consent, so you need to make up your mind as to whether you feel happy with moving forward with a sweep or not. Depending on where you are based, this intervention can be offered prior to 40 weeks. Some women decide that they do not want any form of intervention at all; others feel that when faced with a possible chemical induction they would rather give the 'less invasive' option a try. Neither decision is right or wrong. This is your body, your birth and your baby. You have to do what feels right for you.

Inductions
If you are a low-risk mother experiencing low-risk pregnancy, it should be unusual to be offered an induction prior to your estimated due date. However, an increase in blood pressure, a prolonged start to labour after waters breaking, or concern for baby's growth rate, are all reasons that this may occur.

Whilst hospital policy will vary from region to region, once a woman reaches 41 +1 in the UK, particularly if she is birthing in a hospital, her caregiver will want to start discussing the option of being booked in for an induction.

Inductions are an extremely emotive subject within the birth world. I interviewed Virginia Howes, author of *The Baby's Coming* and independent midwife who has supported thousands of couples during pregnancy and birth. Virginia firmly believes unless there are specific medical reasons, women should absolutely not be induced. You can listen to Virginia's interview by visiting www.thecalmbirthschool.com/bookbonuses

My personal view on induction is a little more grey simply because I don't want any of the mothers who go through The Calm Birth School course to be totally petrified if they end up having to go down the route of induction. Having said that, choosing to be induced should be treated as a huge decision, as a medically managed birth means that your body doesn't produce its own oxytocin and endorphins in the same way as a birth that is able to progress naturally. The consequence of this is that many of the benefits we talk about in terms of being able to enjoy a comfortable and more relaxed birth are more difficult to experience. Often the process makes birth much more intense. However, anecdotally, hypnobirthing mothers who have trained themselves to deeply relax on demand and are able to go with the sensations they experience, as opposed to fighting against their body, are in the best position to still create a more positive induction experience. Go back to being clear on the risks and benefits of moving forward with any procedure during pregnancy and birth and make your decisions based on what feels right for you and your family.

If you do decide to move forward with an induction, my key piece of advice is to build up your bank of endorphins and get your natural oxytocin level as high as possible before moving into the hospital environment. Put the following three priorities at the top of your induction checklist:

1) Feel happy. Once you know you're going to be induced, you also know that within a couple of days maximum you are going to be holding your baby in your arms for the very first time. Woo hoo!
2) Listen to the fear release MP3. Acknowledge any fears or anxiety about being induced you might be holding onto and then let them go.
3) Start building up your bank of endorphins. Do some lovely things prior to going into the hospital.

For example, if your birth partner is your life partner too, go on a lovely walk and soak in the fact it will only ever be you two together as a twosome ever again. If this is your second or third baby, celebrate the changes this new baby will bring into your life.

Other suggestions:
- Have a delicious meal
- Go for a massage
- Do lots of things that make you laugh
- Be intimate with each other

These will all help to build your bank of natural endorphins and contribute to you feeling more relaxed and at ease before your induction begins. Remember, should you start surging regularly, you can always request that you continue to labour normally without further stimulation. (I recommend putting this in your birth preferences.)

Please remember you will still be able to use your Calm Birth School techniques should you go in for an induction. In fact, it's very important that you do. Because you will know your date for induction, you will have a head start! So in terms of your preparation, use this to your advantage.

The Induction Process

Most women will have been offered up to two cervical sweeps prior to experiencing a medically managed labour. Often a gel or a pessary is inserted into the vagina. These contain artificial prostaglandins which help to ripen the cervix and stimulate uterine waves. If this is unsuccessful you will be asked if you are open to having your waters manually broken. This is usually done with a small hook that looks similar to a crochet hook. If this isn't enough to get things moving you will be offered syntocin or picotin depending on where you are in the world, which is a synthetic form of oxytocin, administered via an intravenous drip.

Many women find the pressure waves stimulated by these artificial drugs are much more intense than the body's own natural waves. When induced with this drip, the body does not respond by producing endorphins in the same way as when a woman goes into labour naturally, which is why your own preparation, should you be having an induction, is so crucial.

It's not for me!
If however, you decide induction is not for you, this too is a valid choice, *you do not have to move forward with any medical care*, including being induced. This can often feel like a difficult decision to come to, which can leave you feeling very isolated unless you are under the care of an independent or private midwife, because you will be going against the status quo, and your care providers may be against waiting for nature to take its course.

Often women are told they run the risk of their bodies not functioning as effectively if they allow their birth to continue past the 42 week period, alongside risking the health of their babies.

This is something no mother-to-be wants to hear, and many tend to proceed with the intervention despite feeling uncomfortable with the choice, or as if they do not have alternative options. However, the risks when stated in this fashion do not present a full picture, which means mums-to-be are unable to make an informed decision. I strongly recommend that if you are thinking about induction and are not sure whether it is right, you visit www.aims.org.uk (Associations for Improvements in Maternity Services). This charity provides robust statistical, evidence-based research which is extremely accessible for those of us without medical backgrounds. This will allow you to assess whether induction is the right decision for you or not.

As I have said many times before, hypnobirthing is about understanding what you can control and letting go of what you can't. In terms of medical interventions, you are always the person in charge. No one can prevent you or 'not allow' you to do anything: and you always have the right to make a decision based on informed consent. You have the right to choose not to move forward with an induction. You have the right to decline vaginal examinations. You can choose to birth at home after a previous caesarean. You have the right to ask for evidence to support the advice you are given, and there are many sources of information that you can seek out independently to help you to make an informed choice about how you would like to proceed with your pregnancy and your birth. When faced with any non-urgent medical matters, use your BRAIN system (Benefits, Risks, Alternatives, Instincts, Nothing). Ask questions and do the research so you look back and know you had the best birth experience for you on the day.

If you decide an induction is not for you, but you would like to feel even more at ease with your decision, you can request additional monitoring of both you and your baby. You can ask for the amount of fluid around your baby to be checked. You can ask for your baby's positioning to be checked too.

Bear in mind that we were not designed to stay pregnant. And barring special circumstances, our body gets it spot on. So if you are a low-risk mother, enjoying a healthy pregnancy and all the signs point to a healthy baby, many people would argue your baby hasn't been born yet because your baby is simply not ready. Your body will kick into action as it was designed to when your baby is ready.

Special Circumstances

We have already discussed birth preferences and their importance, particularly in situations when birth does not go to plan. Whilst it is extremely important for you to focus on creating the calm, positive birth you desire, sometimes even with the best planning in the world, special circumstances may mean you will have to look at your B and C options for birth preferences.

Sometimes women who have ended up utilising relief for the intensity of their surges or perhaps met an unplanned scenario, have felt like there has been some kind of failure on their or their body's part. I want to emphasise that no-one can fail at birth and your aim is to enjoy an experience that you want to look back on with joy in your heart whether your baby arrives vaginally, breech, with the help of forceps, an epidural or via caesarean.

All of these experiences can be as magical as the next when you feel clear on why certain precautions or actions are taking place, and when it feels right for the health of you and your baby. The aim is *always* for a positive birth, *not* a perfect one.

Alongside birth not playing out exactly as we anticipated, for what can be a multitude of non-medical reasons it's also important for you to be mindful of the clear medical reasons why you should seek out support during your pregnancy or labour should you notice any of the following:

1) Feeling your surges start prior to 37 weeks. Although it's very common to experience Braxton Hicks, a tightening of your tummy without much intensity in the latter stages of pregnancy, if you are surging and noticing intensity before the 37 week period you should call your care team straight away.
2) If you notice a change in your baby's regular patterns of movement. Many people mistakenly believe that baby's movement will slow down towards the end of your pregnancy. This isn't the case.
3) A prolonged period of time from your waters releasing and the onset of labour. I include this tentatively and urge you to do some of your own research around this, as many women are offered antibiotics or inductions on the basis of safety for both mother and child when in fact, the evidence to support the normal course of action for most medical care providers suggests that waiting to see if labour progresses normally without any intervention actually poses significantly less risk than what is often presented. Check out this article:
   http://midwifethinking.com/2010/09/10/pre-labour-rupture-of-membranes-impatience-and-risk/
4) Noticing that your membranes are discoloured or have a strong odour to them once they have released. This could be meconium (baby's poo) which can be a design of some distress for baby, while an odour might indicate a urinary tract infection.
5) Regular and severe headaches.

6) A very high temperature or fever.
7) Extreme vomiting or diarrhoea.
8) Blurred vision or dizziness particularly if combined with substantial swelling of hands, face, ankles or feet.
9) Heavy bleeding. Whilst it's normal to experience spotting towards the end of term, any heavy bleeding should immediately be checked out by your care team.

As always, don't forget to tune into and trust your instinct; if something doesn't feel right, even if you can't put your finger on it, receiving reassurance from your care team that everything looks okay and there's nothing to be concerned about is a great endorphin producer which will always be a good thing for both you and your baby.

Chapter 13: How to Help Your Baby Along Naturally

When your baby is almost ready to make an appearance, sometimes it is possible to give them a little encouragement and help them on their way. Whilst I don't believe any of these suggestions will be any help if baby's not ready, they will be useful if all baby needs is a little nudge in the right direction to make his or her appearance.

Sex

Sex and/or physical intimacy is the biggest and the best thing you can do to nudge things along. When we get intimate, particularly when we are climaxing, we produce huge amounts of oxytocin. So touching, kissing and making love are all fantastic ways to help move things along.

Reflexology

Reflexology is great at around 36/37 weeks. A reflexologist skilled in maternity reflexology will stimulate certain pressure points within the body that are known to stimulate labour.

Acupuncture

A series of sessions from around 36/37 weeks can be very helpful. Some practitioners offer specific packages for avoiding induction too. Acupuncture originates from Chinese medicine and the methods used work with the body's energy systems and meridians, moving a woman's energy to help baby make an appearance.

Acupressure
Works in a similar way to acupuncture, without the needles, and focuses on stimulating labour by applying pressure to points across the body.

Being physically active
Walking up stairs and steep hills is great, because you tend to be in a forward leaning position, which helps get your baby to put more pressure on the cervix in order to help stimulate the uterus into action.

Dates
Eating six dates a day from 36 weeks onwards has been shown to increase the likelihood of spontaneous labour and reduce the chances of intervention. Get the medjools in!

Nipple stimulation
Using either your hands or a breast pump will help produce the hormone oxytocin, which you will definitely know by this point is great for stimulating birth.

Spicy food
Any spicy food that might make you want to go to toilet, basically. Kick it up a notch. So, if you're used to eating a

chicken madras you need to go for a Vindaloo or a Phal, whereas if you're normally a korma girl than a madras would be fine.

(… Oh, and did I mention sex?!)

Emma's Story

I was sent for a growth scan at 36+ weeks and baby was measuring fine but fluid was on the low side, so we had to go back for monitoring every couple of days. All was fine, but I was then told that since I was 37 weeks, why wait? I was originally supposed to be having a home birth so this would have been a big change for me and I wanted to wait. We asked for another scan and daily monitoring.

The second scan showed that the fluid level had reduced even more so we agreed a date for induction that worked for us in terms of arranging childcare for my first child and co-ordinating when our doula would be available. I ended up being induced at 38 +5. I did lots of research about what would and could happen, researching pain relief and telling the midwives that I would ask if I needed something. The midwives were very positive about trying to make it as natural as possible. We packed some fairy lights and my affirmation posters and decorated the room.

I had two pessaries. The first did nothing and my cervix was not ready at all so we walked for miles around the hospital grounds. The second pessary was put in at 3:30pm and I started to get mild cramps about an hour later. I was examined at 6pm and my cervix had just started to dilate. Things started to get a bit more intense so I put the birth rehearsal MP3 on repeat and tuned into my breathing.

My doula arrived just after 9pm and the midwife examined me and said I was 4cm, which to be honest still felt like I had so far to go. I refocused, and every time I had a surge I would just breathe and focus on waves crashing against the sand.

At some point I asked for gas and air, and a bit later on I felt I needed an epidural mainly because I was really tired and there wasn't a break. The anaesthetist arrived after what felt like forever; in hindsight, I was in transition. It was about midnight at this point. I was finding it very hard to listen to what he was saying and feeling an intense pressure. The midwife insisted on examining me whilst I was having several surges and said I was 8cm, another surge I was 9cm and then one more and I was 10cm. Baby Ophelia Isabel was born at 1:24am with minimal tearing and no epidural. I was vaguely aware of the anaesthetist leaving the room at some point!

From start to finish, the whole process was 7.5 hours, much more intense than my first 36-hour directed purple pushing forceps labour, but so satisfying to know that I could deliver completely naturally. I was much more in the hypnobirthing zone. I think in a way because I was induced I knew it was coming, and I started my breathing straight away rather than being too far gone like last time. During my first labour, my doula had to work hard to talk me through each surge. This time, I was chatting, then I would have a surge, and then pick up conversation again. It was amazing and the scripts and general calmness of The Calm Birth School really helped.

Chapter 14:
Signs of Early Labour

The Seven Signs of Early Labour

Early labour is known as the 'latent phase'. Contrary to popular belief, not all of us will 'just know' when we have gone into labour. Some women will, but if you find yourself second-guessing all of the usual aches and twinges as you approach or pass your guess date, do not fear. You are not alone. To help determine if this really is your moment, here are seven helpful hints of what to look out for.

1. *Cramping*
A pretty universal description for the start of labour is a period-like, cramping feeling, often felt in the abdominals. If this is your only sign, it probably means that things are beginning to move along, and it's understandable that you may be uncertain about whether it really is time. Doubt often arises if the sensations are irregular, with a complete lack of intensity and no other accompanying signs. As you've seen from Laura's story above, sometimes early labour takes a while.

Some women experience weak cramps for days or even weeks before they move into established labour; for others things progress more quickly. If you experience cramping, look for consistency in the frequency of the sensations to indicate if early labour is starting. If you are destined to experience a long lead up to labour, infrequent cramping sensations can be frustrating and confusing. This is can be referred to as a long latent phase, or practice labour.

A good way to deal with the uncertainty is to practice the acceptance you will be using whilst birthing. Don't wish for things to hurry up, or think about when you are going to know for sure. If you are able to, go about your everyday business, staying as active as possible. If you can, get outside and go for a walk. Bear in mind when you are on the move that you might want to stay within an hour of your home, just in case. Continue to reassure yourself with the knowledge that your body is getting prepared. Sooner or later you *will* be holding your baby in your arms, so relax and go with the flow.

2. *Waters Releasing*
This is one of the first signs that spring to mind for most people when we talk about indications for the onset of labour. Once again, a note about language: rather than talking about waters 'breaking' I encourage you to describe them as releasing. Nothing is broken during labour. It's all good.

What you might be surprised to learn is that the release of your waters is not a given. You won't necessarily experience a big gush of liquid like you've probably seen in the movies. Sometimes this does happen, and it's not unusual for some women to choose to sleep on towels, or have some type of protective covering for the mattress, just in case. Some women will gush for hours on end, if this is you there is nothing to worry; you're just releasing a lot of water.

However, many women find themselves unsure as to whether their waters have released at all, as they only feel dampness. I call this *trickling*. If you're a trickler, you're more likely to find yourself wondering whether you've wet yourself rather than if your waters have released. Sometimes you'll hear an audible pop when they waters release. Again this is nothing to worry about.

In some cases your care providers will want you to travel to your place of birth to confirm whether your waters have released or not, and many will ask you to put on a sanitary towel so they can check for moisture and confirm the waters are clear and baby hasn't passed any meconium. Meconium is the equivalent of baby doing a poo in the amniotic sac and will turn your clear waters a shade of brown. It can sometimes — but certainly not always — be a sign that baby is in distress.

Other care providers are happy for you to stay at home and wait for you to travel until your surges are established. Speak to whoever you're working with in advance and find out what the protocol is so you can make appropriate plans for the big day.

3. *Losing Your Mucous Plug*
Your mucous plug sits in the neck of your womb and as your cervix begins to get wider, this mucousy membrane is discharged. Losing your mucous plug is a fantastic sign that your body is preparing for birth. However, be aware that this

can happen up to two or sometimes even three weeks before you go into labour.

Your mucous plug can be clear and quite snotty looking, but it can also be a bit pink in colour and sometimes look bloody. The loss of the mucous plug can sometimes be referred to as a "bloody show," because the mucous becomes blood stained. This is totally normal and nothing to worry about.

4. *Diarrhoea*
We love a loose stool at The Calm Birth School! A bit of diarrhoea before or whilst you are experiencing early cramping sensations is a big green light that your body is expelling everything it needs to, making the path of descent as easy as possible for your baby. Bring on the poo!

5. *Feeling or Being Nauseous*
Nausea is another fantastic sign your body is doing what it needs to do and is clearing your digestive system so your body is in the optimal state for birthing.

6. *Excessive Nesting*
Most women want to welcome their baby into a lovely home that feels fresh and clean, which is why you might get sucked into all of those drab tasks you've been putting off for weeks. However, if you notice yourself verging on obsessive, this could be a positive sign things are due to start happening. Think of it as you instinctively preparing your nest.

7. *Spotting*

A little bit of light spotting – often partnered with the dislodging of the mucous plug (but not always) – is fine, normal and a sign your body is preparing for your labour and birth. If you notice any blood prior to 36 weeks or feel it is too heavy, it is always worth checking in with your medical care providers.

Bonus Sign: Intuition
You might have female intuition going on and sense you need to stay close to home on the day your labour begins. For example, it's not uncommon to hear of women saying even though they had no signs at all, they just knew that they should stay close to home that day.

What to Do When Labour Starts

During The Night
If your labour starts at night, rest, relax and conserve as much of your precious energy as possible. You don't want to feel tired just as your body and baby need you to feel energised. In the same vein, if your birth partner is with you and you go into labour at night, let your partner sleep if the sensations you are experiencing aren't very intense and you don't need support at this time. You'll want your partner to be as alert and supportive as possible at the exact time you need them to be. Whilst there isn't any doubt that you've got the biggest job on the day, being a birth partner is a tiring job – even if all they appear to do is just sit and hold your hand.

Being in the early stages of labour when you would normally be asleep is the only time I would ever recommend you lying down. However, you can still do this with a slightly elevated back, by propping yourself up with cushions, or lie down on your side to create a more optimal position for baby.

During The Day
If you are in early labour during the day, stay active but more importantly, always listen to your body and rest when you need to, in order to ensure you conserve your energy for when you most need it. When you feel a surge during early labour, stop moving and focus on your breath before continuing on with your activities. Keep a 'business as usual' mindset. Stay active and keep your body as upright and forward as possible. This aids the natural process of working with gravity, helping to encourage your baby to stay in, or move into, the optimal position for birth.

Sitting down, bouncing on your birth ball and laying on your left side can all be great resting positions. One of your birth partner's roles will be to gently remind you to switch positions every 45 minutes or so. If they notice you have been static for two long, encouraging your to move will ensure you do everything you can to encourage baby to move down your birth path optimally.

If any part of you feels apprehensive instead of excited – and this counts during the daytime too – acknowledge the fears, practice your Calm Birth School breathing and release and let go of anything that is causing you anxiety. This will create the best foundation for a calm and comfortable birth.

Natasha's Story

I want to say a BIG thank you to The Calm Birth School and everyone in the Facebook group that offered advice and support during my pregnancy. Thanks to TCBS I was able to have the birth experience I had imagined. My son Buddy Jude Middleditch was born on 10th September and I'm so in love with him! The breathing techniques, MP3s and TCBS wisdom helped me to feel confident and prepared for anything.

I was overdue and an induction was pending but the day before it I went in to labour. I was able to stay at home and work through my surges and then go to our local birth centre. I found the affirmations incredibly helpful and had them on repeat. I wrote some of the positive affirmations down on some cards and added some things that I knew would keep me going and my husband read them to me. This really helped; hearing his voice made me feel safe and got me to relax. I think this helped my husband to come up with things to say when I was struggling.

When I was in the birth pool I remember him telling me to think of the baby's tiny hands and feet and to imagine the fun we will all have together. The other thing that helped me during each surge was to imagine waves coming in and out. During the final stages I imagined I was a lion on the shore and to push the waves back out I'd roar. This may sound strange but it really helped me to maintain my breathing and focus.

TCBS also helped me to cope with an infection I had after the birth. I am so grateful for all the invaluable advice and tools that TCBS gave me and could write pages about it. Thank you so much! xxx

Chapter 15: Positions for Labour

When it comes to finding the best positions for birth, I urge you to follow your own natural birthing instincts and go with what feels comfortable. There are no rules as such, but there are certainly positions that can be more or less helpful for creating more positive birth experiences.

Active Birthing: Sitting/Squatting, or on Your Hands and Knees

Active birthing was made popular in the eighties through antenatal educator and birth pioneer Janet Balaskas. After completing lots of her own research into birth in different cultures, it became increasingly clear that far from being strapped down to beds and often anaesthetised as was common in the West ,in other cultures women often kept moving whilst they were labouring until they felt ready and would then proceed to squat during the birthing phase. A 2012 <u>study</u> by Gupta, Hofmeyr and Shehmar finally provided supporting evidence that women who remained active and upright during birth had a significant decrease in interventions.

Active birthing allows the pelvis to open, providing baby with more space to travel through the birth path. There is also evidence that these type of positions lead to shorter births.

Keeping Mobile During Active Labour

The following positions will help you stay mobile and as relaxed as possible, in as optimal a position as possible, during labour.

- Sit down on a toilet with your torso facing the cistern so you can rest your arms on the back of the toilet. This position can offer great relief alongside helping you open up your pelvis.
- Sitting astride a chair, with your torso facing the back rest, can have the same effect as sitting on the toilet.
- Use your partner to hold onto them around their neck, whilst gently swaying or circling the hips.
- Sit on a birthing stool or chair.

Lying on your back whilst you're labouring will _never_ help you birth your baby more quickly or comfortably. If you find yourself wanting to lie down on your back, discuss with your partner how they can help you stay mindful of moving around – without them sounding like they are ordering you about. Give yourself a limit of half an hour and then move into one of your favourite upright and forward positions.

Please check out Katy Appleton from Apple Yoga's bonus video on prenatal yoga and different types of positions. www.thecalmbirthschool.com/bookbonuses

Irene's story

My baby boy Damiano is now nearly 9 weeks old and it's taken me until now to put this experience, which I will cherish for the rest of my life, into words. First I want to thank the moment I decided to take that card at the Baby Show in London. I have always been a fan of relaxation, breathing techniques and so on, but I couldn't imagine that they could be applied to birth. Thank you Suzy for the wonderful work you do and the support that TCBS gives to mums-to-be.

Damiano was born at 1:28 am on Tuesday 21st April and he brought immense joy and happiness, after 23 hours of labour! It started on Monday 20th April at 2.45am when I was woken up with a light cramping sensation. At first I didn't think they were surges and I kept going to the toilet and going back to bed. I was 38 weeks +5 and in my mind I thought that I had another couple of weeks ahead. I couldn't sleep any longer and at 4am I woke my husband up and told him what I was going through. He started timing the surges and although they were coming quite regularly, they weren't strong at this point. We kept timing them until around 8am when we decided to call the midwife, since by this time the surges were regular - coming every four to five minutes - and lasting for a minute each time.

The midwife visited us at home. I was only 1cm dilated so she told us to keep doing what we were doing and to wait until the surges were stronger before calling her back. In the following hours I took a nice warm bath, used the ball in all possible positions, listened to music) in particular to Adrift by Christopher Lloyd, which accompanied us till the delivery of the baby) and kept breathing, breathing, breathing.

At 3pm we called the midwife again as the surges were stronger and quite close to each other. She told us to go to the hospital, as I should have been at the right stage by then. We called the taxi, grabbed everything needed and went to St Thomas' hospital. In the meantime my sister arrived too, and alongside my husband she was a great birthing companion. During the taxi journey I kept my eyes shut, visualising my baby and breathing at each surge with my favourite affirmation in mind: my surges cannot be stronger than me, because they are me!

We were given our room at the hospital (with breathtaking views of the Houses of Parliament by the way) and a midwife visited me telling me that I was 2cm dilated. Only 2cm after so many hours?! I couldn't believe it, but in true calm birth style, I stayed cool. We were told that we couldn't stay as the hospital policy stated that 2cm wasn't dilated enough. We were so upset and tried to persuade her. In the end she told us that we could stay another hour after which she would have assessed me again.

You can imagine that we weren't exactly happy with the possibility of being sent back home, and I think my body knew it because all the sudden there was a gush as my waters released. We called the midwife in, who said that it was better for us to go back home into our familiar environment. Another gush! It was as if my body was telling me, "No you can't go anywhere; you are ready!" By this time, the surges were regular and were getting stronger and stronger.

A new midwife visited me and I was 6cm dilated this time. Wow: from 2cm to 6cm in less than two hours!. She said, "Don't worry... you are not going anywhere!" Once we were settled into the room, I had a nice bath to try and relax after the 'go home' fear and then spent time on the birthing ball and on the floor, with my partner breathing with me and my sister giving me a nice back massage. The midwife even helped with aromatherapy!

I asked whether I could use the birthing pool and it was available. I kept feeling the need to push and at each surge I tried to push and breathe the baby out, but after a while my surges were not strong enough and couldn't manage to push him or her out (we didn't know the baby's sex). We spent 40 minutes in the pool, but as the baby's heartbeat was slowing down, I had to come out. It was a shame; the warmth of the pool was lovely, but got me too relaxed!

Once back in an upright position, my surges became nice and strong again. We tried to go back to my room and a strong surge came, which nearly made me have my baby in the hospital corridor! That would have been fun. The midwives were all ready with the mattress on the floor and I couldn't care less where I was by that point. Anyway, we managed to make it back to the room and once I was on the bed, I gave three pushes and my lovely baby was out!

He was the best and sweetest thing I have ever seen in my life. After some minutes, the cord was clamped and I cut the cord myself (my husband is not keen on blood and squishy things that much). Cutting our cord gave me a strong happy sensation. Damiano was placed in my arms and I delivered the placenta naturally within 10 minutes, so the lovely journey of motherhood started skin to skin with my baby! During all those hours of labour I had no pain relief, no gas and air, nothing: just me and the support of my husband, my sister and the two lovely midwives, Charlotte and Suzanne, who helped deliver Damiano.

A massive thank you to TCBS for such useful tools too, without which the whole experience wouldn't have been as positive and empowering as it was. The only tiny downside were the many stitches I needed as Damiano came out with his hand next to the head, which left me with 2nd degree tears. During the stitching I accepted gas and air, which put a big smile on my face!

The morning after, the new midwife on duty asked me how it went and my answer was, "I ENJOYED IT!" She looked at me like I was from another planet. "Well," I continued, "I wouldn't do it again tomorrow but it has been a beautiful, positive and almost painless experience."

In the end we spent two nights in the hospital because I lost lots of blood and I had to be checked the following day, plus Damiano wasn't latching properly so they wanted to monitor us properly.

Nearly 9 weeks have passed as I write this and Damiano has been such a calm baby. I'm loving this precious time with him. I love being a mum. Thank you TCBS.

Chapter 16: Active or Established Labour

What Active Labour Feels Like

Depending on where you are in the world, active labour is considered to begin from 5cm dilation onwards. Sometimes women report they feel more intensity once established labour begins, some say as soon as they felt their first surge it was all systems go, and others report no difference between the early stages of labour or after they have surpassed the 5cm mark. Remember: everybody experiences birth differently.

If you feel immense intensity straight away, rather than the gentle build up many expect, remember you have all of the tools you need to access a deep state of relaxation quickly. By simply using your Calm Birth School breathing (in for four and out for seven) followed by your wave breathing (in through the nose for seven and out through the mouth for seven) you won't need half an hour to enter a deep state of relaxation. You will be able to access this state quickly and easily.

However, in order for this to come instinctively to you, you *do* need to have been regularly practicing your breathing techniques in situations non-conducive to relaxation. Don't wait until it's late at night and you're all warm and cozy in bed to practice your breathing. Go for it when you're in the middle of chaos, being bumped about on public transport, or wanting to explode at your birth partner because they have said the wrong thing… again. When this happens, your body will become used to responding with relaxation once you step into the less familiar territory of giving birth.

When to Travel to Your Care Provider

If you're birthing outside of your home, the closer you are to established labour of five centimetres dilated or more, the less likely it is you will need any medical intervention. For first time mothers, I recommend that unless you are going to feel safer or more comfortable travelling to your place of birth earlier, stay at home until you experience four waves in ten minutes, each lasting for about a minute, consistently for two hours. For those of you who have done this before, three waves in ten minutes, lasting between 45 seconds to a minute is a good rule of thumb.

If your birth partner is with you, let them do the timing and measuring the frequency of your waves.

Some women prefer not to use a clock to time their waves for a valid reason: whenever we engage with something that requires us to use our rational mind, it is more difficult to distance ourselves from the sensations we are experiencing in our bodies. Having your birth partner in control of the clock can help if you're going down this route. Having some kind of signal or trigger word so they know when to start and stop the timer can also minimise chatting if you are the type of person who prefers to focus all of their energy inwards when they are birthing. Maybe you'll be extremely engaged and talkative in-between surges and until you get to the day, you simply don't know which camp you will fall into, so it's useful to have a plan.

An alternative method for measuring how far dilated you are comes from the book *The Art of Midwifery* by Hilary Marland and utilises the fact the body is diverting more blood away from our legs and feet towards our uterus in order to increase its efficiency whilst we are birthing. When a woman is between 1cm and 2cm dilated, the feet and ankles are colder; once she reaches between 4cm and 5cm, the calves are colder; and then at full dilation of 10cm, the legs are cold from the knee downwards. Clever right? Be aware that this is not a technique you can use if you have been getting in and out of the shower or bath, since the warm water will change your body temperature.

If you're birthing at home, talk to your care provider about when would be the best time during the process for you to contact them before they come to you.

Whenever you experience a surge, use your Calm Birth School breathing techniques to eliminate tension in the body. Then move into your wave breathing, aiming for equal length inhalations and exhalations. Ideally, your body will be as limp as possible, like a rag doll, so there is no resistance in your mind or your body.

When you call your care providers, often they will want to talk directly to mum, as they like to listen to how you are breathing and to assess how the waves are affecting your ability to speak. Let them know you are a hypnobirthing mum as it's likely you will be managing the normal stress and intensity of labour much more effectively than they will anticipate, leading to some care providers to conclude you are not as far along as you really are. Make sure you monitor the frequency of surges and communicate this to them.

Chapter 17:
Coping With Distractions

How to Deal With Distractions

If you're choosing to birth in your own environment, it's much easier to feel comfortable about managing the potential distractions around you. When you know you will be moving from your home to a birthing centre or a hospital, there are inevitably more distractions you will encounter along the way, many of which you will not be able to do anything about. I want to reassure you that outside distractions don't need to be the end of your positive birth experience. You are a master at filtering out distractions in your everyday life and we will be tapping into this awesome skill of yours in the lead up to and on the big day itself.

The brain processes 400 billion pieces of information every *second*, of which we are aware of around 2,000. Then we filter out the things we deem unimportant or irrelevant to what is going on in our surroundings. This filtering method stops us from going insane. Thank you, nature! This mechanism is precisely why you don't need to get your knickers in a twist about being distracted. It is also the exact same mechanism I want you to tap into when you're birthing. You can start practicing it right now. In fact you already are, but perhaps aren't fully conscious of it yet.

Your guide for when to consciously tune into this filter and when not to is really simple. If you are faced with a situation that is creating tension or irritation within you and it's something you are able to do something about, (a dripping tap or a snoring partner for example), then *do something about it*. Take charge; these situations can be remedied quickly and easily.

If however you're faced with a tension-inducing situation that you do not have the power to change, choose instead to focus your attention inward and start to watch and engage with your breath using the Calm Birth School breathing techniques (surprise, surprise!). Allow the distraction to take you even deeper into a state of relaxation. You can even say to yourself, "The sound of [insert distraction] helps me drift deeper and deeper into a beautiful state of relaxation." A great time to play with this is when your baby is having a kick and a stretch while you are doing your daily relaxation exercises. Any time there is intrusive noise or something happening within your line of sight, use it as an opportunity to consciously connect to your breath.

This will be easier or harder to do, depending on the distraction, but the more you practice, the easier it will become. A great example is when someone else's child is crying on a plane. Some people will be there wishing they were in the peace and quiet of Business Class, while others will be able to fall asleep in the midst of chaos, appearing completely undisturbed. Your opportunity to practice might come on a bus or train, while waiting for an appointment or perhaps when you're at a family gathering. The aim, by the time you go into labour, is for you to be a person who can choose exactly when and where you want to focus your attention, and who can breathe through *anything*.

By making the conscious decision to focus on your breath and allow the distraction to help you go deeper into relaxation, you can change the whole experience of the noises around you with ease. This is a skill that takes practice, but you *can* do it. The trick is to start, so I encourage you to make a start today.

Massage

To help manage the internal distractions – i.e. your waves – take advantage of the positive power of touch.

Touch is an important part of intimacy between a couple. Touching can help us to relax, and feel connected and safe. All of these emotions are great for producing endorphins and oxytocin, which we now know are the ones we want for a quicker, more comfortable labour (as opposed to cortisol and adrenaline). I recommend that a couple makes time to really tune into this during pregnancy by enjoying daily massage with each other — and, when I say with each other, I mean the birth partner massaging mum! Some people will enjoy a firm pressure, but the massage I recommend at The Calm Birth School involves a very light touch.

I love this method because the light touch often creates a tingly feeling within the body, similar to that "barely there" feeling. Taking 15 minutes out of your day is not only a great way to build on your opportunity for prenatal bonding, but the touch of the birth partner becomes a cue or an anchor for mum to become even more relaxed every time this type of touch occurs, which can be extremely valuable during labour. However, it is also worth noting that, even after taking the time to enjoy this fantastic ritual during pregnancy, when you are labouring you might not want anyone within an inch of you! Some mothers have reported they needed to be alone so they went and birthed in the bathroom. That's also absolutely fine. Massage is an extra tool you can call upon should you want it on the day and it will be infinitely more powerful if you take the time during your pregnancy to really connect a sense of calm to your partner's touch.

How to do it
The technique is simple: your partner or birth partner will use their fingertips or the backs of the finger nails in a gentle upward motion, stroking and moving the fingers up the back and across your shoulders, almost creating a T shape. Repeat and then repeat some more.

You can also bring these soothing strokes up and down the arms, across the chest and nipples, up the neck, behind the ears and into the hair. This is definitely a useful tool to draw upon if labour slows down.

The Dial Down Method

Another great method for distancing yourself from both internal and external distractions is using the Dial Down Method, which is a self-hypnosis tool. This is a great tool to use whenever you notice yourself feeling tense or stressed and you have a little bit of time on your hands. It's the perfect technique to draw upon if you are on public transport, or when travelling.

How to do it

Imagine a large dial, with the numbers one to ten going around the perimeter. As you imagine the dial, see the gauge hovering at number ten. Then take a deep breath in, imagine breathing in calm and as you exhale, imagine breathing out tension. On the completion of that exhalation, see the gauge move to the number nine. Again, inhale calm and exhale tension, then see the gauge move down to eight. Do this all the way to zero and you should notice how much more relaxed your mind and body feel. When you get to zero you can continue to focus your attention on the breath, or imagine relaxing in your favourite place. Play with it and follow your gut instinct.

You can amend this technique and visualise a big temperature gauge, a sliding scale of musical notes or a thermometer – whatever makes the most sense to you. Most people start practicing this with their eyes closed, but as soon as you become familiar with it, eyes open is equally as effective. If you're ever in the office, particularly if you are in one of those meetings – you know the ones I'm talking about – and you need to get out of your own head for a bit, this is a great trick to have up your sleeve.

When TCBS mums first start using this technique, it doesn't always come naturally. Many of the techniques I've shared may feel difficult to get your head around at first. Please persevere and be kind to yourself. It is the same when you learn any new skill – it takes time and a lot of repetition to master any new practice to a point where you can do on autopilot. Think about when you learned to ride a bike or drive a car, or a toddler trying to find their walking legs: how many times will you see them pull themselves up, only to topple over? It's all part of the learning process, and it takes time, effort and perseverance. It's exactly the same with the skills you are learning throughout this book too, which is why I recommend starting *now*, not a couple of weeks before your guess date!

Time Distortion

Many times, hypnobirthing mums have talked about the quickening of the passage of time when they are birthing. This is a noticeable indication of being in a trance state. It's similar to when you've been engrossed in a great book you have just no idea where the time went. You can have the exact same experience during birth of time working with or against you, depending on how you are feeling in the moment.

Time flies for a woman who feels relaxed and calm and looks forward to each surge as an indication each sensation brings her one step closer to meeting her baby. Contrast this with a woman who dreads the every surge and wonders when it's finally going to be over and how long labour is going to last. Every minute feels like ten... or more.

When you feel good, your birth partner can capitalise on this sense of time passing more quickly, by suggesting every twenty minutes feels like five. Whilst it might feel silly reading this right now, remember that when you are birthing you are going to be in a trance state and therefore more open to suggestion, so use that in your favour.

Laura's Story

On Monday morning we finally got to meet our **gorgeous** *daughter Imogen Sophia Kill. I was very lucky and had an officially short active labour but I wanted to share the full story to show how TCBS got me to that point and how my much longer prelude to labour wouldn't have been so calm without the techniques this course taught me.*

I was very anxious about labour from the moment I found out I was pregnant and started looking into hypnobirthing straight away. I had heard about TCBS indirectly through a social media post by Tom from McFly (cringe - I'm a mega fan!) and signed up for the course when I was 20 weeks.

At first I studied religiously and posted affirmations around my house. I also worked hard on educating my family and fiancé so that they were supportive and not dismissive of my desire to use hypnobirthing. I believe the support from others was key to the course's success for me. Don't get me wrong - my fiancé wasn't begging to watch a video or listen to an MP3 every night (which we did in fact do at the end of my pregnancy), but he was on board with the concept and I was so pleased.

After I signed up I did my best to practice where I could but self-doubt often crept in that I wasn't doing enough. If you're reading this thinking the same thing about yourself, have faith: whatever you are doing will be going in! This was where my fiancé's support was key; he would remind me of the techniques when I had doubts during pregnancy and during labour.

My surges started at 9pm on a Friday but had disappeared by the Saturday morning. They returned on Saturday lunchtime but again were gone by Sunday morning, only to return again Sunday lunchtime (phew!) at which point they then increased in intensity and frequency.

Emotionally these few days were quite draining. I was constantly thinking, 'Is this going to be it?' and I had plenty of time for self-doubt and panic to set in. However, instead of letting this happen, I watched numerous calm birth videos that had been shared through the TCBS group (the support you can get from the Facebook group is amazing) and I listened to my MP3s regularly over these few days to keep me calm.

In all my preparation, I had not considered that I might be in early labour for a few days. Each time I felt a surge I would practice my breathing (always being reminded by my fiancé to do it), and a random affirmation would pop into my head! I also used a TENS machine to help with the pain relief. Finally on Sunday evening at about 12:30am I was having three surges per 10 minute window so called the midwife-led unit and they agreed that I should go in.

I was 3cm dilated at 1am so they gave me two hours before they checked me again, offering me pain relief tablets which I declined. By 4am I was at 4cms and my options for pain relief opened up. During the waiting to get from 3cm to 4cm & making decisions about how my labour progressed I reverted to the BRAIN acronym and felt empowered to be making those decisions and not sheep dipped into having a medically managed birth.

Once I was at 4cm I moved onto gas and air then had a wonderful aromatherapy massage from the midwife. Things progressed pretty quickly from there. I went from 4cm to delivering my baby in the water within 3 hours with just 20 minutes of pushing!

I have to be super honest here and say I was not silent or serene like some of the incredible women we see in the videos, but I was incredibly empowered and I felt okay being very primal while I gave birth. I learned that that was more than okay and it clearly resonated with me on the day!

Overall my biggest credit to TCBS would be keeping me calm while in early labour and throughout my pregnancy. It educates you to be a woman who is fully in control of the decisions you make. Your labour may not go the way you hoped and you might have limited choices, but you do have choices all the same and you should feel comfortable in making them without fear there. Thanks for the education Suzy and for helping me have a genuinely positive and empowering birth experience. I couldn't have done it without you.

With love and gratitude from me, my fiancé & our precious little girl.

Chapter 18: What Happens Once You're With Your Care Provider?

A Moment in Time

Once there is space for you on the labour ward or in your room – if you're not birthing at home – most of the time you will be offered a vaginal examination. Note the use of the word *offered*. A vaginal examination, as with all procedures offered during birth, is provided on informed consent, so it is your right to decline a vaginal examination should you wish to.

Why do some women decline this? Well, some mums-to-be want to have a completely undisturbed birth and believe baby will arrive whenever he or she is ready. Their perspective is that a vaginal examination does nothing to aid that process. It's also recognised that a VE only measures a moment in time. So, if you go to the hospital, are examined and a well-meaning midwife says, "Oh, you're only 2cm," the use of words 'only,' plus the fact that you are 2cm can be hugely deflating. If you do opt for an examination, remember that it *is* just a moment in time, that labour isn't linear and that things can progress very quickly.

I cannot emphasise this enough: if you opt for a vaginal examination – and many women do – and you are not progressing as quickly as you would like, relax and do as many oxytocin and endorphin producing activities as possible.

Partners: this really is your time to shine and remind your birthing beauty that this is just a moment in time and let them know they are doing amazingly. Just because it's taken ten hours to reach three centimetres, it does *not* mean it's going to take another ten hours to get to six. When a mother is relaxed and birthing as actively as possible, it is completely possible and not uncommon for mum to go from three centimetres to eight in an hour! If you find that you haven't progressed at the rate you would have liked, know that it's totally normal to feel frustrated and concerned about how much longer things are going to take and then consciously choose to release the expectation or the wish for things to be different. Let go of any tension, doubt or anxiety that may be trying to creep in and simply be accepting of the moment.

This is not easy to do if you haven't been practicing breathing and sending relaxation to the different parts of your body, both at home and in less conducive places for relaxation, such as on public transport. So make sure you practice in advance!

The Slow Down

Whether you're travelling to a hospital or birthing at home, it's very common at one point or another during your labour for things to slow down. Depending on what stage you decide to leave home, if you do at all, a slowdown occurs because you are leaving your cozy and familiar home environment. This is normal and all part of the evolutionary armoury that aims to keep us and our babies safe from harm while our subconscious assesses the reasons we have moved and assesses whether it feels good to continue with the labour.

If your body and baby decide to take a little break, The Calm Birth School recommends:

1. Breathing techniques
You will have practiced these so many times by this point that your muscle memory associates breathing with feeling good, being calm and at ease, which acts as a signal for you to produce more endorphins and oxytocin.

2. Touch
Remember the power of positive touch, whether that be a little bit of massage, stroking or holding. If you feel open to being held or touched, give your birth partner the green light, if they have not already taken their cue.

3. Walking

Walk around. Walk upstairs while leaning forward to help baby put more pressure on your cervix. Walking sideways upstairs can also help – think like a crab!

4. Visualisation
Visualise your baby moving down the birth path and how good it's going to feel to be holding them in your arms, or listen to your Birth Rehearsal MP3 – all roads lead to oxytocin.

5. Nipple stimulation
Is great if things slow down as the stimulation of the nipples impacts the body in a similar way to when a baby is suckling, causing the body to produce oxytocin to help promote bonding and attachment. Of course oxytocin production will increase the efficiency of your labour.

6. Laughter
Watch anything that makes you laugh. Download it onto your tablet beforehand so you've got easy access to it. We laugh when we're relaxed and when we're relaxed we birth babies more quickly, easily and comfortably.

Trust that your body and your baby know what to do and things will pick up again at exactly the right time. The best thing you can do is relax.

While I think films (hilarious ones) are a great idea, mobile phones, laptops or anything that will engage the rational, analytical part of your brain should be banned during birth if you are looking to create a quicker and more comfortable birthing experience.

I recommend no technology once you start labouring: no tweeting, Facebook or texting your friends. The more you disengage from the left side of your brain – the part responsible for critical analysis – the easier you will find it to connect with the sensations in your body, rather than analyse them from a cognitive distance.

It's also worth thinking about whether you want to let anyone know once you have gone into labour, because if you are going to experience a longer labour, the pressure of having people texting you or your birth partner to find out what is happening can also create unnecessary stress, pressure or analytical engagement.

Working With Your Care Provider During Labour

Ensure you have at least two copies of your birth preferences with you, so if there is a change of shift when you're birthing, your new care provider can get up to speed.

As we have discussed, who you have in the birthing room is hugely important. and your birth partner should take on the responsibility for managing your environment. Highlight this section for your birth partners to read.

It is the birth partner's role to be fully tuned in to mum, so they can see how she is feeling and talk to the people who are working for you. Please remember they are working for you. If there is any hint your practitioners are not supportive, or there is a personality clash, your partner needs to take the lead and in a calm but firm way ask for you to be cared for by someone else. It is not your responsibility to be concerned about how that is managed, but it is your partner's responsibility to hold and protect the space, so that you have an optimal birthing environment, which is full of peace, love and support. Then you will be able to make the decisions that are right for you and your family.

Remember you will only have this experience once. So you have to make the choice about whether it's best to be more concerned with ensuring that you create the most positive birth experience for you and your family or worry about potentially offending someone who is unable to support you in the way that you need in that moment... I know what I'd choose.

Decide beforehand whether you would like your birth partner to act as the middle man between you and your care providers when it comes to making decisions about the birth, or whether you would like to take the lead. As always there are no right or wrongs when it comes to this, but if you are looking to birth more comfortably and quickly, the less you engage in conversation, the better.

Many hospitals will want to see progression from mum of about 1cm dilation per hour. Unfortunately, because women are not machines, we often do not work like this. Once again it's important for the birth partner to be tuned in to how mum is feeling so you can alleviate any pressure you may encounter from people trying to speed things up.

If you find yourself in a situation where people are trying to hurry things along, use the BRAIN system and ask very specific questions. This will provide you with specific answers and allow you to make informed and personal choices about the course of action that is right for you. Always bring any suggestions about proceeding back to the specific indications at that moment.

Example Questions

"What specific evidence suggests that there will be a risk of X taking place?"

"What specific signs right now indicate that waiting for the next thirty minutes would be harmful to mum or baby?"

"If we chose to wait another 45 minutes, what are the possible risks, and what evidence do you have that those risks are applicable to me and my baby?"

Always be specific.

Language

By this point I hope you are being mindful with the type of language you are using. I've talked at length throughout this book about using different terminology and being very mindful of the kind of thoughts you think. Unfortunately you can't control the words used by others. That said, it won't do you any harm to include on your birth preferences you would like your care providers to talk in terms of *comfort levels* and *waves* as opposed to *pain* and *contractions*.

Certain language used within the medical profession can sounds very loaded and can be upsetting or even distressing to hear during labour. Prepare yourself in advance by making a note of the terms below and making a decision about what you and your birth partner will do if you notice this language during your birth.

For example, if someone in your space is using negative language, it is totally acceptable for the birth partner to respectfully ask the care provider to speak with them directly first, quietly and even outside the room if possible. The birthing environment should be kept as emotionally safe and relaxed as possible.

Failure to progress

While I am on a mission to ban the word *failure* from birth, this is unfortunately all-too-common a phrase used in birthing environments, which can lead the most well-prepared woman to feel as though she's not doing things properly.

Here's the truth: *you cannot fail at birth.* If you happen to hear this when you're labouring, remember that you always have options about whether you would like to follow the lead of your body, or if you would appreciate some assistance. Once again go back to your BRAIN and assess what you feel is going to be right for you.

Incompetent contractions

Woah! Incompetent is *definitely* another phrase that should be banned from birthing rooms. There is nothing incompetent about your body and sometimes, for good reasons, your baby will stop moving down the birth path and you will require assistance. It might simply mean that you need to move around, change positions and take the pressure off in order to allow the oxytocin to do the job it was designed to do.

With any decisions you make about how involved you would like your care providers to be, *remember that you have a choice.*

If you opt for support to move things along a little, when you are able to make these decisions while feeling calm and in control, because you were able to ask specific questions and receive specific answers, then you remove any ambiguity. This hugely contributes to you being able to look back on your experience knowing you had the right birth for you, regardless of how your baby enters into the world. That can only be a positive thing.

Things to be mindful of during the active phase include: changing positions and keeping as active as possible; using positive touch and words for comfort and reassurance to make your birthing environment – particularly if you are not birthing at home – as lovely and as nest-like as possible; having your birth partner hold the birthing space so it feels calm, private and safe.

Chapter 19: Second Stage Labour

Your are considered to be fully dilated at 10cm. This is sometimes referred to as the down phase or second stage of labour and it means your baby is close to being ready to emerge. There are three main factors to be aware of, which I'll talk you through now.

A Pause

At this point, just as during early labour, the body may take a natural break as you and your baby instinctively prepare for this new phase. This is not the same as the slow down, and it's nothing to worry about or hurry along as long as baby's heartbeat is strong and consistent. Sometimes this pause will be ten minutes, and I've even heard stories of the break lasting a few hours. If you are birthing at home with an independent or private midwife, it is far more likely they will feel comfortable going with the body's lead at this point. If you are birthing in a more medicalised environment, it's more likely that your care team will want to move things along for you if the break is deemed to be too long. Should you be faced with this scenario, remember to use the BRAIN system to determine what course of action is going to be best for you and your family and to make sure you are clear as to why any suggestions are being made.

Sometimes instead of a pause, you might notice an increase of intensity or frequency in the waves you are experiencing, or perhaps a change in the sensation of the surges. Everyone's experience is different.

Pooing

One of the big fears some women have is pooing during labour. Shit does happen (sometimes) but remember that it's all for our own good: once we move into the second phase, the body's primary focus is to work as efficiently as it can, thus expelling everything that may inhibit baby from moving along the birth path as easily as possible. As baby moves past the bowel, any waste is emptied so that the birth path is as unobstructed as possible, which in turn makes baby's descent easier.

So, should you start to feel an overwhelming urge to do a poo, be happy: it means you are in your second stage and so close to meeting your baby. Sometimes you might feel the urge to do a poo but not actually end up doing one. This is great too. It simply means that baby is moving down the birth path but that you don't have any waste to expel.

If you feel the need to go to poo once you are fully dilated, resist the urge to jump up and go to the toilet if you can, and move your attention back to your breath. It's time to begin breathing your baby down. If you are not participating in vaginal examinations, trust yourself to know that things are changing and you are moving into the second phase. Some people may call this 'pushing phase'. For some women it will feel as though the body is taking over and for others it will be a more conscious decision to change the way they are breathing with their surges. At this point I invite you to choose to focus your energy and your energy and your breath downwards, working *with* your body and assisting your baby's journey.

Feeling Nauseous

Another great indication you are moving into the second phase is being or feeling sick. If you find yourself projectile vomiting across the room after a substantial period of labour, with this accompanied by either an increase in the frequency or intensity of your surges. Alternatively, you may notice your surges dying off as your body pauses before you prepare to cross the final hurdle (there's no one direct route). It's all positive!

Pushing and Breathing

Breathing Baby Down
When we watch people giving birth on TV, the second stage is what we usually associate with lots of forced pushing and care providers shouting and coaching the birthing mother. Some of you may want the support of your midwives, and some of you may prefer quiet to let the body do what it innately knows how to do. As always, nature thinks of everything and, in just the same way as for every other birthing mammal, there is no

need for the majority of women to force their baby out. The body has a unique mechanism for doing this on our behalf, called the *natural expulsive reflex*. You will already be familiar with it from using it every day when you go for a poo.

When you are in this downward phase, the best thing you can do to aid this process is imagine sending your breath and energy down into the ground, past your baby, around your uterus and into the floor. This allows your body, pelvis and vagina to be as relaxed and open as possible for a quicker and more comfortable birth.

Some women experience an overwhelming urge to push or bear down. If this is you, I advise you to go for it. Listen to your body, do not resist it, even if 15 minutes previously you were told you were 'only' 6cm dilated. As I said earlier, vaginal examinations only measure a moment in time and things can change very quickly. Trust your body.

So what's the difference between pushing or bearing down if your body is telling you to do so, and forced or coached pushing? In a word: tension.

Tension
What I'd like you to do, either on the toilet or sitting in your chair right now, is to pretend that you're pushing out a poo. What do you notice? You should notice the muscles around your sphincter contract and tighten. Which is the exact opposite of the action we are looking for with a calm and positive birth for both you and your baby. When you are able to breathe your baby down and work with your body, your

baby tends to make their entrance in a calmer, less explosive way, which also helps to keep the perineum intact. Bonus.

Having said that, for a multitude of reasons, some women opt to force push their babies into the world. As with every piece of advice or insight I offer throughout this book, if this is a conscious decision, coming from you, then it's all good. You cannot get it wrong. If you are faced with a situation that dictates that forced pushing is the best way forward and it feels right for you, then it's right. I really want you to hear me on this: *you cannot get this wrong.* Just go with what feels right for you on the day.

Birth Breathing

Use this technique when you are experiencing a wave or surge when fully dilated.

How to do it
The best way to aid the natural expulsive reflex is to work with the breath in a similar way as when you are wave breathing. The main difference is that you place *all* of the emphasis on the exhalation. The out breath needs to be very long and very deep. It can be useful to use a visualisation to accompany the out breath, anything that reminds you of the importance of staying open, relaxed and moving downwards.

Some women use the words "open, relaxed or release," others will think about there being no resistance, or imagine a flower opening, or will picture something significant to them that helps keep the idea of openness in their mind.

It can be really helpful to work with noise when you are getting to this stage. Although, some will feel equally comfortable working with the breath alone, others will want to hum, shout, groan or even moo. If it hasn't already, it can get incredibly primal at this stage. This is nothing to be fearful of for either you or your birth partner. No resistance is the main aim of the game. If you want to howl, just howl! Whatever you instinctively want to do is all good - seriously. The only thing to be mindful of is to use the noise and the energy to send your power back down to your baby and your uterus so they can finish the job.

When to use it
The best place to practice this technique is when you are having a poo. If you are at home, hum when you're on the loo so you start to feel more comfortable and familiar with directing your sound and energy down in that way.

This is great if you're suffering with constipation too. It won't shift everything immediately but by applying patience, the humming and the breathing will see your natural expulsive reflex start to get things moving much more quickly and comfortably.

Chapter 20: Transition and Crowning

I Can't Do This!

There often comes a moment during birth when a woman thinks, "I can't do this anymore!"

The second phase usually lasts a maximum of two hours – although this will varies from woman to woman. Sometimes, right before baby is about to emerge, a woman may feel overwhelmed, like she can't do it anymore and wants to give up. This is called the transition stage. If this happens to you, this is an amazing sign for you and your birth partner that you are now within spitting distance (excuse the analogy!) of meeting your baby.

This is the time for your birth partner to remind you of what an amazing woman you are and let you know you have reached the final hurdle. This period of wanting to give up or feeling like it's all too much right before the prize doesn't happen to everyone. However, if it does happen, take peace in the knowledge it is only likely to last for around 15 minutes, and you really are going to be meeting your baby any moment.

Once transition has passed should you experience it at all the last moments of birth are all working with your body to breathe baby into the world. Just as baby is about to emerge some women experiencing an intense tingling sensation. You will notice this just as baby's head is about to emerge. This is your baby stretching and pulling on the vaginal walls. This intensity only lasts for a short time and is often followed by a numbness, so aim to stay relaxed. Sometimes the sensation of your baby's head can come as a shock – as crazy as it may sound don't be tempted to close your leg (it does happen!). Focus on your birth breathing and waiting for the next surge to breathe out your baby's head fully, which should then be followed by their body, in just two to three surges.

Charlotte's Story

Throughout my pregnancy I struggled with varied physical and mental stresses and pains. I had numerous hospital visits with kidney stones, urine infections and bleeding, which, when accompanied by running two businesses, working 15 hour days, 7 days a week, plus teaching dance classes and renovating a house, made me only too pleased to take a day off work one Friday. I enjoyed a mum-to-be spa day which was a baby shower gift from my lovely friends, and I treated my mum to treatments too as she had been a fantastic support to me throughout my pregnancy.

We had watched all the Calm Birth School videos together and discussed how I was going to have an ultra calm and happy birthing experience. I had been having cramping for about 6 days and she would remind me to remove my frown and stay positive. So after a relaxing day off work, and a 3 hour afternoon sleep (!) I came home feeling really calm and relaxed. That evening at 11pm, as I sat on the sofa with partner Steffen, my waters released. I was so excited, bouncing around with so much joy singing "We are having a baby! Baby is coming, he's coming so soon!"

I think it was this happiness that set off the endorphins and oxytocin and things went from 0 to 60 from that point. Within 30 minutes the surges had started but very soon I realised that there weren't any breaks in between them.

For about an hour I struggled to move, get dressed or talk. I knew I was in established labour: the feelings were very intense and painful and honestly, there was a point where I thought, "I just cant do this." I knew that I needed to get to the hospital; maybe then I could get some pain relief.

The birthing centre that we had chosen was closed due to staffing shortages, so we made a quick change of plan and went straight to the local hospital. When we arrived I was wheeled in because by this point I couldn't even walk. I knew I was close to pushing but because I was silent with my eyes shut the midwives didn't believe I was in established labour and left me in a waiting room, with horrible bright lighting.

I was determined to stay strong, positive and above all, calm! Finally I was examined and the midwife was shocked to be able to see my baby's head! With no time left to fill up a birthing pool as planned, I was rushed to a labour room where I got on all fours and began to push. I told Steffen to put my Calm Birth School playlist on and get the lights down low. I focused on the positive affirmations whilst mum and Steffen took it in turns to rub my middle back, which was hurting a lot.

I would highly recommend essential oils like clary sage, which really helped speed things up whilst relax me. It took about 30-40 minutes for me to breathe my baby out, literally without any pain relief or even a sound. It was an amazing experience and being really positive and fearless about giving birth is what I think helped me the most: calm mind, calm body, calm baby, calm birth!

It is so important to remember this and I can't thank TCBS enough. Meeting you and learning from you helped to make my birthing experience a truly magical and special event and it is now my job to spread the message that giving birth can be amazing and very positive.

Lots of love from a very happy family.

Chapter 21: Your Baby's Here!

Even though many a wise parent will tell you that the real work begins once baby arrives, there's a little bit in between giving birth and parenting starting that is worth thinking about and preparing for.

Those moments where you hold your baby for the first time are not only magical because you are getting to look into each other's eyes, but also because you start to close the complex physiological and emotional loop of birth.

Cleaning Your Baby

If your baby is born on dry land as opposed to in the water you will notice white coating on your baby's skin. This is called vernix. Some mothers prefer baby to be completely wiped down before enjoying skin to skin with their child; others prefer things to be left *au naturel*. The choice is up to you. However, some people argue that babies are not born dirty and the rush to clean them just isn't necessary; vernix acts as an antiseptic and a moisturiser for baby's skin with the antiseptic properties protecting baby from a whole host of infections, which is amazing! Again however, if you prefer to have baby cleaned before skin-to-skin contact, this is absolutely your choice.

Delayed Cord Clamping

Delayed cord clamping is when the umbilical cord is not immediately cut from baby. Some parents are happy to leave it for two minutes before cutting, whilst others prefer to wait until the cord has stopped pulsating completely before cutting so they know baby has everything they need from the placenta.

One of the things you can research is whether delayed cord clamping will be suitable for you or not. As you've read in some of the stories from TCBS mums throughout this book, the idea of delayed cord clamping might seem perfect for you, until you actually go into labour, when circumstances can change very quickly.

If this is of interest, please do your own research into the benefits and drawbacks of this course of action, and be mindful your choice is very important to include on your birth preferences sheet.

Birthing Your Placenta

It is worth researching whether you would like to have a managed third stage when birthing your placenta, or whether you would prefer to let the process take its natural course. If you have a managed third stage, you will receive an injection of synthetic oxytocin to stimulate the uterus into surging. This often means the third stage is over within half an hour. A managed third stage is often recommended, as it tends to reduce the number of women experiencing a large loss of blood after birthing. One of the disadvantages of taking a synthetic drug to speed the process up is that it increases the chances of retaining parts of your placenta, which can lead to infection and sickness. Once again it's important to do your own research so you can decide which option feels right.

Skin to Skin

Skin to skin is one of nature's most awesome design features, and creating the space to allow this to happen immediately after birth has a *long* list of benefits! So, barring special circumstances, this is a fantastic way to welcome your baby into the world.

Skin to skin contact with your baby stimulates oxytocin, another amazing design feature of the birthing process. It turbo charges the bonding process, helping you want to protect and nurture your baby immediately. However, it is important to say, even with the oodles of oxytocin coursing around your veins, some mums take a little while to connect with their baby fully, and if this is you, that's okay and a normal response too.

The oxytocin not only helps with attachment bonding, but begins stimulating the uterus so that within the hour you will have birthed your placenta.

When your baby lies directly on your bare chest – and on your partner's, too – they start to colonise with your friendly bacteria, which is amazing because it means alongside the anti-bodies in your milk (if you are choosing to breastfeed), you begin protecting your little one from any unsavoury bugs and germs that may be lurking.

One of the by-products of oxytocin production is a warm, fuzzy feeling. The warmth helps to start regulating baby's body temperature and reduces their cortisol levels (the stress hormone). It also triggers your milk ducts so you can begin feeding baby right away.

If, due to special circumstances, you are unable to move forward with immediate skin to skin, but it's something you don't want to miss out on – particularly if your baby is preterm – you can always ask your care providers about Kangaroo Care. This is where mum gets to carry baby around on her chest, like a little joey, which again helps to form attachment, help with breastfeeding and importantly colonise with your friendly bacteria to help protect against illness and infection.

Breastfeeding

Breastfeeding, as with birth, is an individual journey that for some women comes very naturally, and for others can be extremely challenging.

Whilst I don't go into any details about the practicalities in this book please do check out www.thecalmbirthschool.

com/courses for the Infant Feeding Workshop. What I would like to touch upon is the emotional side of feeding because, next to lack of sleep, it's often one of the things that takes new mums most by surprise. I often say that it's not until you have your own child that you appreciate why so many mothers think of themselves as 'feeders.' Ensuring your child has enough nourishment is solely down to you if you choose to breastfeed. If you everything does not fall into place naturally around breastfeeding, the feelings of guilt and shame at not being able to feed your baby in the way you had envisaged can be hugely overwhelming.

The evidence shows that statistically women who seek out support and guidance before their baby arrives tend to have a more positive breastfeeding experience. Women who seek out assistance from a lactation consultant as early as possible if things are not going as well as anticipated are far more likely to breastfeed for longer. Don't suffer in silence, or wait for six weeks before acknowledging that help would be good. Needing help does not make you a failure.

If after seeking out support, you still feel as though breastfeeding isn't for you, please do not beat yourself up. The happiest children have happy mothers - not strictly because they were breastfed. Sometimes, for a multitude of reasons, you may choose to stop breastfeeding. Please hear that this doesn't make you a bad person or a bad mother. Educate and empower yourself and remember, it's your body, your baby, your decision and nobody else's business.

For a more in depth help with breastfeeding please check out Infant Feeding workshop here: www.thecalmbirthschool.com/courses

Conclusion

I hope as you reach the end of this book a quiet confidence has been awakened within you, or that perhaps the inner strength that you have been silently drawing upon now has its own voice that you feel very comfortable sharing with the world.

Hypnobirthing isn't about producing perfect, textbook births. If only it were that easy. However, when as a woman you understand what you can control, and you let go of what you can't, you can step into creating the most positive birth experience, regardless of the way your baby chooses to enter into the world.

This is *your* body, *your* baby and *your* unique experience. You are a warrior woman, a goddess who was born to give birth. You now know that birth is a natural and normal event, not a medical condition.

You know that being able to embrace and accept what is going on in your body, rather than resist it, can help you to manage the sensations of birth far more effectively.

You understand that although your emotional state will have a huge impact on the way that you experience birth, the birth environment, your care providers and your birth partner also have extremely important roles, and all of them should be there to support you fully. You have the right to ask for what you want and need.

You now have a toolkit of specific techniques that will help you to remain feeling calm and at ease throughout your labour and birth.

So what's next?

No one can force you to do anything you don't want to do, but in order to get the most out of this book, I suggest that you go right back to the beginning and read it again, highlighting all the salient points. Yes, seriously. Remember everything I've said about preparation. Never underestimate its value.

There really is no secret to giving birth, other than to be present in any given moment and to accept what is going on with your body. The nub is, for most of us out there, that it takes time and practice to be able to tap into the acceptance that you need on the day. So hear me when I say that I believe in you and you have got this, *and* make sure you put the time in, do your practice and watch how you are able to take whatever birth sends your way with ease.

Next, set dedicated TCBS time aside in your calendar to practice the breathing techniques and visualisations. Don't simply set the intention to practice: by putting actual appointments into your calendar, you'll make the appointments with yourself "real" and will be much more likely to show up for them. - especially if you've got a full life already. Remember you can download The Calm Birth School practice guide at: www.thecalmbirthschool. com/bonuses

In closing I want you to know, it's so not about birth for me, it's about life. This is about you. I want you to know that you count and your voice is important.

While giving birth should never define a woman, the process of fully connecting with your inner voice, strength and wisdom that you will learn to exercise, stretch, pull and embody both during your pregnancy and birth will provide you with an opportunity to celebrate and own your magnificence as a woman. As a human being, who is not only a living, breathing, walking miracle herself, but as someone who also creates miracles.

As you gain confidence in communicating your needs, wishes and desires for you and your unborn baby and begin to understand the power that you contain is the power needed to conquer worlds, whether these worlds are within the private sanctuary of your home, or breaking down glass ceilings in the boardroom. This path (whilst not the only way) is a path to your power in life.

It starts with know what you want and telling other people and then looking your fear in the face. Staring at it head on and seeing it melt away as you trust in yourself, your body and your baby to lean into your strength.

When you are able to navigate your birth in a way that leaves you feeling like a lioness. It changes things.

It changes *you*.

You know you can do anything.

You can do *ANYTHING*.

You have got this!

All the love
Suzy xo

Post Script

One more birth story? Oh, go on then.

Hayley's Story

Today is my due date. However, 6 days ago our little Huxley Isaac decided to come into the world.

My waters broke at 11:30pm on Sunday, when I called out our midwife to check me. She suggested trying going back to bed and said it could be a while before anything happened. (I'd been having really strong Braxton Hicks for a week.) I managed to doze off until about 4:30am then decided to get up; the surges were mild and 10 minutes apart so I called my mum at 5:30 am to come over to help get our four-year-old to school.

I then spent the next few hours trying to doze off, walking around the house and having small snacks, all the while welcoming the surges. I found walking and standing up for them was the only way. We had lavender burning, candles and plinky plonk music as my other half calls it! It was super relaxing; my mum felt as though she was in a spa!

I did some walking to the end of the garden and back again as I didn't want to be far from the house but the surges when we were timing them (which wasn't all the time) were ranging from four minutes apart and then back to eight or nine.

We called the midwives again at 11:30am. They told me to just relax, not do anything let my body do its thing and if I wanted, to have a bath. They did mention that they thought that being second baby, it would have been a bit further along but not to worry. I did say that I had a bit of an ache down below but not a pushing feeling which confused them, so one midwife advised me to have feel myself in the bath to see if I could feel anything. After getting in the bath and feeling what I thought might have been the head, and 3 close strong surges later, the midwives luckily were already en route.

The next couple of hours out of the bath went so fast. I had the affirmations on while I rocked holding onto my partner or mum, I found "ahhh-ing" sometimes on the out breath of wave breathing helped. The pool was only just filled in time; I got in at 3:18pm and only 3 surges later at 3:34pm he was here!

I am so thrilled with our calm positive birth experience which despite being definitely uncomfortable at times towards the end was also in a way euphoric. This birth was 16 hours from start to finish compared with 30 hours my first, and I had only 3 surges where I let go and birthed him compared with 2 and half hours of forced pushing with my first. We were in the comfort of our home and tucked up in bed within an hour!

Thank you to The Calm Birth School for everything I've learned, changing mine, my mum's and hopefully others' perception of birth and the support of everyone on this group.

Further Reading

Childbirth Without Fear	Grantly Dick-Read
Active Birth	Janet Balaskas
Men, Love & Birth	Mark Harris
Guide to Childbirth	Ina May Gaskin
Hypnobirthing	Marie Mongan
Birth Reborn	Michel Odent
Gentle Birth, Gentle Mothering	Sarah J. Buckley
The Expectant Dad's Handbook	Dean Beaumont

Additional Support

The Calm Birth School Video Program

Whilst I offer a very limited number of face-to-face consultations per year for hypnobirthing, one way to experience the equivalent of a private class in the comfort of your own home is to invest in the Calm Birth School video program. The benefits of this include:

- You are able to control and direct your learning by participating in the classes as many times as you want or need, to right up until the day you go into labour
- You can and learn in two different formats - reading and visual/auditory via the video - which many people find suits their learning style
- You might find it easier to engage your birth partner in the video program.

- All students of the video program also get direct access to me through a secret Facebook group separate from the main Calm Birth School Facebook group, where I will personally coach and advise you through any questions, concerns or fears that may arise during your pregnancy journey.

All purchasers of the book are able to purchase the course with a 25% discount using the coupon code TCBS25BOOK

Birth Trauma

If you have experienced a previous traumatic or difficult birth experience and would like to see how additional support could benefit your hypnobirthing journey, I offer in person and Skype hypnotherapy and psychotherapy consultations. Please email info@thecalmbirthschool.com for availability.

Appendix: Breathing Techniques

The Calm Birth School Breathing Technique

TCBS breathing is a simple, powerful technique. The reason it is so effective is because it triggers the body's natural calming reflex, which occurs when the out breath is nearly twice as long as the in breath. During your labour, this technique will help you maintain a deep state of calm and you'll use it in between surges. You will also use it as you feel a surge coming in and once it has subsided.

Ideally, the breath is taken in and out through the nose as opposed to the mouth as this gives you more control over the flow of air. However, please don't stress if you have a cold when you're birthing. Just breathe through your mouth – it will all be okay.

How to do it
Breathe in deeply to the count of four through the nose, imagining you are filling your lungs right to the bottom. As you breathe out, imagine sending the breath down, so it moves around your baby, down your legs into the tips of your toes and then into the floor. Sometimes when you are first practicing this technique, it's useful to put your hands on either side of your waist, so you can feel the rise and fall as you breathe deeply. It really is as simple as that.

When to use it
Use this technique whenever you find yourself feeling stressed, whether it be at work, with your partner, getting on and off public transport – wherever and whenever. This will help you to relax quickly the more that you practice it. In addition, should you be one of the many of ladies who labour quickly, this technique will get you into the birthing zone quickly and easily once you have become accustomed to using it in your everyday life.

If you are the type of person who tends to take life in her stride with very little stress, this doesn't make you exempt from practicing. It's just as important you carve out some practice time for your TCBS breathing too. Ensure you do one set of Calm Birth School breathing in the morning for five minutes, five minutes at lunch time and five minutes again in the evening.

Wave Breathing

Use wave breathing when you are experiencing a wave or surge (otherwise known as a contraction).

How to do it
The central idea of wave breathing is to keep both the inhalation and the exhalation even. Breathe in through the nose to the count of seven and out through the mouth for seven.

The role of this breath is to work with the upward motion of the uterus as it rises, and then to send your breath down to your baby and your womb, while relaxing. Simple!

However, don't be fooled. For you to move instinctively into that space of deep breathing and relaxation when you experience a wave, you need to have practiced it often so that you are used to it.

Please do not worry if you are unable to keep the breath even for a count of seven to start with. Work with what ever feels most comfortable for you. Perhaps you'll start off counting to four and once that feels good extend it to five. The main point is you start to feel comfortable slowing your breathing down and taking control of the flow. This will help you immeasurably during labour and birth.

When to use it
My suggestion is to practice this type of breathing for five full minutes every morning. If that means setting your alarm five minutes earlier – do it. It's such a great way to start your day and will leave you feeling great as well as preparing you for your labour day, when you'll be using it during each surge you experience.

Birth Breathing
Use this technique when you are experiencing a wave or surge when fully dilated.

How to do it
The best way to aid the natural expulsive reflex is to work with the breath in a similar way as when you are wave breathing. The main difference is that you place *all* of the emphasis on the exhalation. The out breath needs to be very long and very deep. It can be useful to use a visualisation to accompany the out breath, anything that reminds you of the importance of staying open, relaxed and moving downwards.

Some women use the words "open, relaxed or release," others will think about there being no resistance, or imagine a flower opening, or will picture something significant to them that helps keep the idea of openness in their mind.

It can be really helpful to work with noise when you are getting to this stage. Although, some will feel equally comfortable working with the breath alone, others will want to hum, shout, groan or even moo. If it hasn't already, it can get incredibly primal at this stage. This is nothing to be fearful of for either you or your birth partner. No resistance is the main aim of the game. If you want to howl, just howl! Whatever you instinctively want to do is all good - seriously. The only thing to be mindful of is to use the noise and the energy to send your power back down to your baby and your uterus so they can finish the job.

When to use it
The best place to practice this technique is when you are having a poo. If you are at home, hum when you're on the loo so you start to feel more comfortable and familiar with directing your sound and energy down in that way.

This is great if you're suffering with constipation too. It won't shift everything immediately but by applying patience, the humming and the breathing will see your natural expulsive reflex start to get things moving much more quickly and comfortably.

As mentioned above, you will use this technique when you are experiencing a wave or surge when fully dilated.

Acknowledgements

The biggest thank you has to go to my long suffering husband, Jerome Ashworth. I am so grateful for all of his support and belief in me. During all of the long nights and weekends he urged me to keep on going and for that I will be forever thankful! My beautiful children, Caesar and Coco, who won't know for a long time what they have inspired, but without them, there would be no Calm Birth School.

Hollie De Cruz, original co-founder and the person who gave TCBS its name. The godmother of hypnobirthing as we recognise it in the West, Mary Mongan. The founders of The Wise Hippo Dany Griffiths and Tamara Ciafinni, who act as a constant inspiration with the work they do to empower women and their families to create calm and positive birth experiences.

Mark Harris, father, midwife, author of Men, Love & Birth for the support he offered The Calm Birth School when it was barely a thought. Katy Appleton of Apple Yoga, Katie Whitehouse from Vital Touch and the Neighbourhood Midwives for their ongoing support.

To all of the couples who have been through The Calm Birth School video program and have kindly shared their journey both within and off of the pages of this book. They are a constant inspiration and motivation for me to do more.

To Stephanie Green for stepping in and making everything she touches of the Calm Birth School look beautiful. And to Elloa Atkinson, for believing in the message of this book.

To the wonder Lisa Rae Fabing for allowing us to use the wonderful image of her son Remo.

Photo credit for front cover image: Love Alda Birth Photographer & Mentor and proud Associate of Nikon.

To all of the childbirth educators out there, who continue to make mothers and babies their priority, I thank you too.

Printed in Great Britain
by Amazon